Morphological Image Processing:
Architecture and VLSI design

Morphological Image Processing: Architecture and VLSI design

PROEFSCHRIFT

ter verkrijging van de graad van doctor aan de Technische Universiteit Delft, op gezag van de Rector Magnificus, prof. drs. P.A. Schenck, in het openbaar te verdedigen ten overstaan van een commissie aangewezen door het College van Dekanen, op zaterdag 28 november 1992 te 10.30 uur.

door

Petrus Paulus Jonker

geboren te Amsterdam,
electrotechnisch ingenieur

Dit proefschrift is goedgekeurd door de promotor Prof. Dr. I.T. Young

CIP-GEGEVENS KONINKLIJKE BIBLIOTHEEK, DEN HAAG

Jonker, Petrus Paulus

Morphological Image Processing: Architecture and VLSI design /

Petrus Paulus Jonker. • Deventer [etc.] : Kluwer

Proefschrift Technische Universiteit Delft. -Met lit. opg.

- Met samenvatting in het Nederlands.

ISBN 90-201-2766-7

Trefw.: computer architectuur / VLSI

Printed in The Netherlands

Ter nagedachtenis aan mijn vader.

Voor mijn moeder,

mijn vrouw Yolande,

en mijn kinderen Filip, Juliëtte, Caspar en Gudo.

Summary

Based on the experiences of past designs and the outcome of recent studies in the comparisons of low-level image processing architectures, a pipelined system for real-time low-image processing has been designed and realized in CMOS technology.
To minimize design pitfalls, a study was performed to the details of the design solutions that have been found in embodiments of the three main architectural groups of image processing; the Square Processor Arrays, the Linear Processor Arrays and the Pipelines. This is reflected in a theoretical model.
As the design is based on bitplane-wise processing of images, research was performed on the principles of Cellular Logic Processing of two dimensional images. A methodology has been developed that is based on the transformation of binary images using sets of Hit-or-Miss masks. This method appeared to be extendable to higher dimensional images. A theoretical model for the generation of break-point conditions in high dimensional images has been developed, and applied up to dimension three.
Based on this result and on the requirements posed by intermediate level tasks, an architecture is proposed for the processing of low- and intermediate-level images in two and three dimensions.

Table of Contents

List of Symbols

Chapter 2:

X	source image
Y	destination image
Z	mask image or masking image
V	value scalar or value vector
S	structuring element in the image domain
$\leftarrow$ or <-	assignment
$\wedge, \vee, \neq, =$	logic and, or, exclusive or, equal
$\overline{x}$	logic inverse of x
$\vec{p}, \vec{q}$	position vectors of image elements
$x_{\vec{p}}$	image element (pixel, voxel)
$x_{\vec{p}+\vec{q}}$	image element in a neighbourhood around $x_{\vec{p}}$
$c_{\vec{q}}$	elements of S; convolution kernel elements
d	transition cost between two nodes
f	evaluation function
$g(n)$	cost from a start node to node n.
$h(n)$	cost estimation from node n to the goal node
d	maximum still optimistic $h(n)$; distance to the goal ignoring the constraints or objects
x_1, x_2	co-ordinate values of an image element in a range image
x_3	value (depth) of an element in a range image

Chapter 4:

X	source set
Y	destination set
S	structuring element set
X	source image
Y	destination image
S	structuring element image; (also: implemented with a mask or mask set)

x	element of x
y	element of Y
s	element of S
$\otimes$	hit-or-miss transformation
$\cong$	inexact neighbourhood match
M_k	3 x 3 neighbourhood around a pixel x_k
$x_k^0, x_k^1, \ldots x_k^7, x_k^8$	East, North East,.....South, Center pixels of M_k
S^s	set of SET masks; element of S
S^r	set of RESET masks; element of S
S_i^s	element of S^s
S_i^r	element of S^r
N	dimension of an image
X_N	N dimensional binary image
X_2	2-dimensional binary image
X_3	3-dimensional binary image
$\vec{O}$	origin of X_N
ε_N	image edge elements (just outside the image)
M_N^n	N-dimensional (hyper-)cubic neighbourhood with size n, with n = 2k+1
M_2^3	3 x 3 neighbourhood in X_2
M_3^3	3 x 3 x 3 neighbourhood in X_3
S_N^d	N-dimensional (hyper-)sphere with radius d
E	number of elements within M_N^n and on S_N^d
V	number of elements within M_N^n and within S_N^d
G_N^d	connectivity between image elements in X_N
d	connectivity distance
$O_{N,\tilde{N}}^d$	basic object
$P_{N,\tilde{N}}^d$	object primitive
$T_{N,\tilde{N}}^d$	tile
D_N^d	layer thickness of a tile
x_N	co-ordinate in X_N
$a_{i,j}$	polynomial coefficient
$S_{N,\tilde{N}}^d$	mask set for G_N^d connected object primitives with intrinsic dimension $\tilde{N}$
ER_N^d	erosion mask (set) for intrinsic dimension $\tilde{N} = N$

$BP^{d}_{N,\tilde{N}}$ break point mask set for intrinsic dimensions $\tilde{N} = 0..N$

$EP^{d}_{N,\tilde{N}}$ end point mask set for intrinsic dimensions $\tilde{N} = 0..N$

Chapter 5:

A (f.g) B generalized matrix product

A (+.*) B arithmetic matrix product

A $(\vee.\wedge)$ B logic matrix product

∂	controller delay
D_{lsp}	delay due to lack of sufficient parallelism in the system
PX_c	local parallel processing claim
PX_a	local parallel processing availability
β_{PX}	overhead factor for the parallel or sequential solution
l:	length of instruction stream for one operation
N	image size
P	number of Processing Elements (PEs)
D_{plt}	program load time
D_{top}	total processing time for one operation
D_{s2i}	sensor to image delay
D_{i2d}	image to display delay
D_{im}	processing time for one image, from sensor to display

1 Introduction.

In the last decade many computer architectures for low-level image processing have been developed. Simultaneously the image processing tasks have been growing and changing, resulting in a larger set of basic operations than the set that was needed ten years ago.
Industrial inspection, robot vision and interactive biomedical image processing are based on more complex algorithms nowadays while a demand for equal or faster system response times remains. Moreover, the gross processing capacity required to fulfill these tasks keeps rising, as satellite data, Printed Circuit Board (PCB) inspection, VLSI circuit mask checking, and document analysis demand larger images, often exceeding a size of 4096 x 4096 image elements (pixels). Image processing in the field of three dimensional image analysis, for example automated microscopy, requires the processing of images with typical sizes of 256 x 256 x 64 volume elements (voxels) and preferably even larger. The massive amount of data in images -with identical datatype-, as well as the data acquisition methods based on raster scanning -such as video frame grabbing-, has naturally led to special computer architectures that fall mainly into two categories, the massively data parallel Single Instruction Multiple Data (SIMD) machines (Flynn 1972) and the data flow pipelines.
Image processing tasks are generally made up of basic image processing operations, which in their turn can be classified in low-level, intermediate-level or high-level image processing operations (Danielsson 1981, Fountain 1986). This is depicted in Figure 1.1.
Within this classification, the low-level or iconic processing stage is the most clearly defined as a class of operations that perform transformations from one image into another. The intermediate-level image processing operations transform image data into symbolic descriptions which then can be used in the high-level image processing stage as basic components for solving certain tasks.
It will be clear that the algorithms that are used in the high-level image processing stage highly depend on the task that has to be performed, for example robot scene analysis, chromosome analysis, crop classification, printed circuit board inspection, etc. The high-level routines specifically use much domain knowledge for the solution of the problem they are designed to solve.

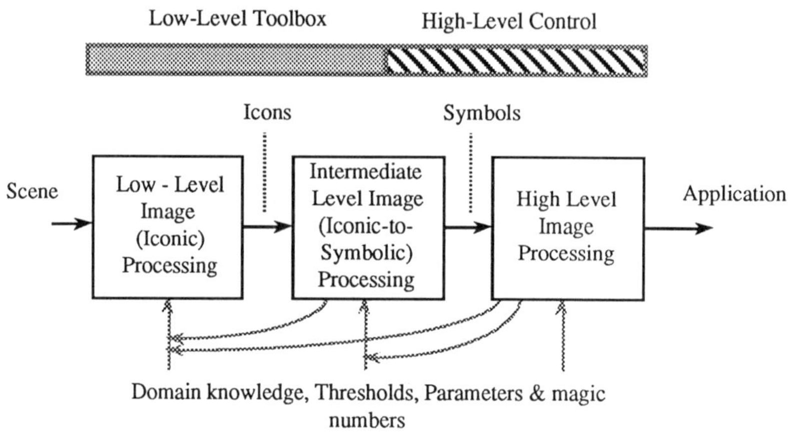

Figure 1.1 Usual distinction in image processing levels.

As a consequence, the intermediate-level algorithms form a class of interface routines between the low-level algorithms and the high-level algorithms. Where the low-level routines can more or less be classified as general purpose, and the high-level routines as special purpose, the intermediate-level routines have often only a limited applicability in other fields of image processing, unless one adapts the algorithm to another field. Where the low-level routines can be considered as data independent transformations with few control needs, the high-level routines can be considered as heavily control bound. This is reflected in the current trend in image processing packages, that are increasingly based on a library of low-level algorithms and a C interpreter for fast prototyping of tasks (Scil-image 1991, Visilog 1988, Imaging-Technology 1990). The evolution from a set of routines with a rudimentary batch-file command language interpreter (TCL-Image 1990, TIM 1989) to a mature interpreter of a common high-level computer language such as C (Kernighan and Ritchie 1988) reflects both the need to create any possible data structure and to interactively make any possible control structure as needed in the intermediate- and high-level routines. Routines, that are based on both an image and a non-image data structure, and that after compilation end up in the library of these image processing packages, can be classified as intermediate-level routines. It can be concluded that only a vague description can be given to the class of intermediate-level image processing routines, as they are mostly application field dependent. As such only the classification "low-level image processing routine" can be justified, and a distinction in a low-level image processing toolbox and high-level programming facilities seems more in place than a pipeline sequence of low, intermediate and high-level routines.

Increasingly common is a form of control on low-level routines that helps to steer the performance of the image processing task. An obvious example is thresholding based on knowledge of the image obtained from a histogram (e.g. Ridler and Calvard 1978). Close observation of many low-level routines teaches us that in most routines hidden knowledge of the image is used. Usually this knowledge is entered in the form of parameter settings such as sizes of filter kernels, etc. High-level routines and sometimes intermediate-level routines are based on this possibility to control the performance of an image processing operation by setting or adapting the parameters of the low-level routine. This feed-back possibility, quite naturally usable in modern image processing packages, must not be neglected nor be hindered by the application of a classification scheme too strictly when designing or discussing new image processing architectures.

In the past decade many special computer architectures have been designed and realized mainly to serve the field of low-level image processing. Characteristic architectures are the Square Processor Array (SPA) or mesh, the Linear Processor Array (LPA) or scanning array, the Pipeline (PL) and the Pyramid (PYR). Due to the massive data parallel character of low-level image processing, solutions are often based either on many small processing elements, constituting a 1-bit Cellular Logic Machine (Lougheed and McCubbrey 1980, Duff 1982, Tanimoto 1986) or on a few, more powerful processing elements such as Digital Signal Processors (Lindskog 1988). Many manufacturers have introduced image processing systems to the market, based on pipeline architectures (e.g. Imaging-Technology 1990), though other promising attempts have been based on a Linear Processor Array architecture, such as the AIS-5000 (Wilson 1988) or on a circular pipeline using the data-flow paradigm (NEC 1985, Iwashita et al. 1986). Finally, specialized component manufacturers have brought dedicated integrated circuits onto the market mainly for filtering and segmentation of images (e.g. LSI-Logic 1989, Plessey 1986).

Due to the advances that the Very Large Scale Integration (VLSI) of digital circuits have offered -such as increasing density of devices, high speed and low power consumption- a new group of architectures emerged that was based on a clock synchronous pipelined approach: The systolic arrays (Kung 1982, 1984). Stimulated by the dynamic character of the basic CMOS circuits, a system design based on MOS VLSI techniques quite naturally led to architectures of devices that operate on streams of data. Most applications are found in the one-dimensional signal processing, although it is claimed that low-level image processing can also be profitably done on these architectures (Webb and Kanade 1986). Until now, however, the research on systolic array-like devices has resulted more in high speed special purpose systems and general purpose arithmetic devices, than in general purpose programmable machines for low-level image processing.

Research in the field of intermediate-level image processing (see e.g. Rosenfeld and Pfaltz 1966, Duff 1986a) has not yet lead to clearly identifiable architectural concepts for intermediate-level supporting machines. Undoubtedly, the Pyramidal group of architectures is a result of the research in the field of multi-resolution image processing (see e.g. Cantoni 1986). Research on local autonomy of processing elements in mesh connected arrays (Fountain 1987, 1988a) and research on data connections to processing elements further away than the nearest neighbour (Hillis 1985) represent steps towards architectures suitable for intermediate-level image processing.

Usually, in high-level level image processing algorithms some possibility for parallel execution remains, albeit this does not mean massive data parallelism. Multi-computer or Multiple Instruction Multiple Data-flow (MIMD) architectures (Flynn 1972) can be fruitfully applied here. As a consequence of the fact that MIMD arrays are not straightforward applicable in the field of low-level image processing, hybrid concepts based on an SIMD or a Pipeline architecture for low-level image processing, and an MIMD architecture for intermediate- or high-level image processing, are proposed (Uhr 1987).

This book reflects a research study on *the design of high speed special computer architectures for low- and intermediate-level image processing and their implementation using VLSI techniques. The basic philosophy of this study is, that all designs are primarily based on morphological image processing.*

Starting from the mathematical morphology (Golay 1969; Preston 1970; Serra 1982, Giardina 1988), cellular logic machines have been designed for two and three-dimensional low-level image processing. As a consequence of the adopted philosophy, low-level greyvalue processing is primarily based on bit-plane wise processing of images, using cellular logic processors. However, this is only carried through to the extent where this approach remains efficient. On top of these low-level designs, facilities for intermediate level image processing have been added.

To allow an efficient design of low-level cellular logic machines for two and three dimensional image processing, morphological image processing has been studied extensively.

The roots of this research can be found in the period after the design of the Delft Image Processor (DIP) (Gerritsen and Aardema 1981, Gerritsen 1982). A need was felt to compare the DIP, with its array-processor like pipelined architecture, to its "nearest neighbour" (in many senses) the CLIP4, an SIMD Square Processor Array (Duff 1982). A research project was formulated for the comparison between Pipeline architectures and Processor Arrays, the results of which are reported in the Ph.D. thesis of Komen (Komen 1990a).

For the sake of this comparison I felt that it would be useful to design a Processing Element (PE) whose architecture would strongly resemble the architecture of the PE of the CLIP4, but based on a data pipeline approach rather then a data parallel approach. For a pipeline system constituted from many PEs, designing a single PE as a special VLSI circuit is profitable because of its repetitive nature. Moreover, experience with the design of VLSI circuits, and insight into the possibilities offered by integration in MOS would be gained by following this approach.

The resulting device, the Cellular Logic Processing Element (CLPE) and the real time systems that can be built with it are described in this thesis. (Real time is used in this context as video speed).

As both designs, the CLIP4 and the CLPE, were based on single bit Processing Elements, hence focussed on cellular logic processing -an embodiment of the larger realm of the mathematical morphology-, special interest was paid to topology preserving thinning or skeletonization of binary images. At that time, and still, many thinning schemes were published that claimed to be "better" than other schemes but did not give much insight into the thinning process itself, and were based on "ad hoc" schemes. The better performance was based on the outcome of their algorithm rather than on a deep understanding of its structure. Starting from the thinning methods of (Hilditch 1969), (Arcelli et al. 1975) and (Beun 1973, used in the DIP; Gerritsen 1982) I started research on breakpixel conditions for the thinning of two-dimensional binary images with the intention that its result should be usable for parallel execution on the CLPE.

As the result of this study showed that a similarity could be established between topology conditions in the image element (pixel) domain and the properties of curves and surfaces in the geometric domain, the use of pursuing this research further to three dimensional images was evident. This was moreover inspired by two ongoing research projects, one in the field of three dimensional biomedical image analysis, and one in the field of vision based robot collision avoidance. In the first application field thinning three dimensional images would be useful, in the latter thinning in even higher dimensions than three could be used for robot path planning. The results of this research on thinning methods for multi-dimensional images are reported in this thesis.

Simultaneously the research on architectures proceeded in the direction of theoretical studies on architectures for intermediate-level image processing and three-dimensional image processing. The aim of this last research topic was to come to a special computer system suitable to perform real-time low-level image processing operations on two-dimensional images, fast low-level processing of three-dimensional images, and to provide services for the flexible programming of intermediate and high-level image processing routines and tasks. Although this last topic is still in a theoretical stage and in progress, the partial conclusions on this subject will also be reported in this thesis.

Chapter 2 of this thesis contains an introduction to low-level and intermediate-level image processing routines and their classification. In a discussion at the end of the chapter some operation primitives will be described. These operation primitives will be used in the architectural designs in later chapters.
In *chapter 3* several special image processing architectures are discussed as well as some general purpose architectures and their estimated ability to perform low-level and intermediate-level operations.
Chapter 4 is devoted to cellular logic processing and conditions for thinning in multi-dimensional images. A theoretical framework is set-up to generate sets of masks that are able to process binary images up to dimension three.
Chapter 5 starts with the introduction of the logic units that were developed in CMOS technology to execute the cellular logic operations as discussed in chapter 4. It is followed by a discussion on the design aspects of low-level real-time image processing systems. As the VLSI design of a programmable processor differs from the design of non-programmable VLSI devices, the third item in this chapter is devoted to the design method that was developed and used for the design of the CLPE. After the design methodology, the design of the CLPE is itself described. Finally, a real-time pipeline system based on CLPE like devices is described. The

system was designed to perform low-level cellular logic operations as well as grey-value operations.

In *chapter 6* a theoretical system design is discussed. This design is a programmable system able to perform low-level image processing operations both on two dimensional and three dimensional images. Moreover, an extension was made to incorporate intermediate and high-level image processing capabilities into the system.

In Chapter *7* the conclusions are presented and the results are discussed.

2 Image processing algorithms.

The design of special computer architectures requires first of all insight in the tasks and algorithms that should preferably run on the proposed machine. Section 2.1 presents a summary of current tasks in three different areas of image processing that should preferably run on this machine.
Section 2.2. starts with a classification scheme for low-level image processing operations followed by examples of each class. The section is closed with a discussion on the architectural requirements of low-level algorithms.
In Section 2.3 four algorithms are discussed that can be characterized as falling into the -vague- class of intermediate level algorithms. The section is concluded with a discussion on the architectural requirements of intermediate-level algorithms.
The chapter ends with section 2.4 in which a final discussion is held on the requirements that image processing tasks and algorithms impose on architectures for image processors.

2.1 Image processing tasks.

As we look at the tasks that are currently performed in image processing a few profiles can be outlined, without pretending to be exhaustive:

2.1.1 Industrial handling and inspection.

Industrial handling and inspection is still a main application field of the classical two dimensional image processing sequence that mainly consists of only low-level routines.
Image acquisition in this field involves both CCD video cameras and line-scan cameras. Gray-value filtering for image enhancement and noise filtering is mostly not necessary, while in an industrial environment the illumination of the scene can be controlled quite good. Segmentation of the well defined objects in the scene from the background is usually done by thresholding the grey-value image into a binary image, followed by post-processing the binary image using morphologic neighbourhood operations. The measurements that are performed are for example the determination of length of lines, sometimes surfaces, perimeter and circularity of objects. Pattern recognition on the result of this image processing sequence is mostly done based on these measured features. It is in this field that most of the

commercially available real-time image processing systems find their applications. Examples are the inspection of Surface Mounted Device (SMD) components before or during placement on printed circuit boards, inspection of parts in robot assembly sequences, inspection during product flow on conveyor belts, inspection during packaging. In this area real-time performance of image processing systems is usually a requirement.

However, more advanced image processing and pattern recognition sequences are also possible. An example in this field is the three dimensional recognition of objects using one two dimensional acquired image and a set of pre-learned two dimensional images of the expected objects (Bart et al. 1990). On the one hand, the processing steps involved after acquiring a two dimensional grey-value image are: Detection of the edges of the objects in the image; skeletonization of the blob-like lines in the binary image into one pixel thick lines; transforming the image from the image element (pixel) domain to the geometrical domain by line following; and building a graph of the obtained line segments. On the other hand a learning set of line segments has been built and optimized. A matching threshold has been determined for the matching of the learned set of line segments with the line segments obtained from an acquired image.

2.1.2 Range imaging and robot vision.

In the field of range imaging or 2.5-dimensional image processing for industrial applications, special image acquisition methods are used to obtain grey-value images in which the grey-value in a certain point in the image does not present the luminosity in that point, but the distance from that point to a reference. Although the range image contains three dimensional information of the boundaries of the objects in the scene, not all three-dimensional information of the objects is available. As the image of a scene is usually acquired from one direction, a range for each object is visible from only that viewpoint. As we are not able to look behind nor inside the objects, the acquired information is not fully three dimensional. From the third dimension half the information is present, and hence the term 2.5-dimensional image is used as more or less equivalent to the terms depth image, height image or range image.

2.5D images are mostly acquired using special sensor systems that are based on triangulation. A well known approach is based on laser lighting and triangulation (Kanade et al 1989, Stuivinga et al. 1989). Examples of applications of these systems in the field of industrial inspection and robotics are: Pallet loading, train rail

inspection, SMD placement, the inspection of blades for electrical razors and robot vision.
Variations on the acquisition method are the use of a scanning laser point-beam, a scanning laser slit-beam, or structured light (Inokuchi et al. 1984, 1986, Vuylsteke 1987, Jonker et al.1990c). Note that conveyor belts, robots and trains can act as scanning devices and therefore be used to omit one of the dimensions X or Y.
An example of an advanced image processing task in the field of 2.5-D image processing would be: Range image acquisition by using a sequence of light pattern projection and image acquisition, decoding of the acquired coded images followed by triangulation, i.e. transforming the camera coordinates of each point in the image (u,v) onto world coordinates of the scene (x,y,z). Note that the parameters of the mapping function to perform this point to point triangulation transformation should be found by a -preferably automatic- calibration procedure.
The obtained depth image can be segmented into objects and background. A method to perform this is the separation of the height landscape of the 2.5D image into surface patches. This can for example be performed by surface fitting, followed by analytically intersecting the surfaces to obtain the object edges in the image. (Besl and Jain 1986, 1988, Schmidt 1989).

Object recognition and the determination of the three dimensional position and orientation (3D-PO) of objects is also possible using a combined stereo vision and graph matching approach. Note that in this approach the number of cameras is not restricted to two. A typical advanced image processing task would be the recognition of objects using inexact matching of graphs (Buurman and Duin 1989). A typical processing sequence consists of high accuracy edge detection, line following, graph building, and graph improvement for two or more cameras. Note that such a sequence can be performed in parallel for each camera. Next, the inexact matching of the graphs of the obtained images to a 2.5D graph, followed by the inexact matching of the 2.5D graph to the 3D graphs of the wire-frame models of candidate objects. These wire-frames can be obtained by extracting information from a CAD system database (see also: Flynn 1991). Note that if this procedure is used for robot vision, the 3D-PO of an object must be obtained with an accuracy better then the robot accuracy (of about 0.2 mm). Imposing accuracy demands on the segmentation phase.

2.1.3 Biomedical image processing.

Biomedical Image Processing is often closely connected with 3D image analysis. Moreover, it is dominated, in contrast with the industrial image processing, by the effect of geometrical distortion of the image due to the sensors and by severe noise due to low intensities. Typical sensor systems are CT-scanners, NMR- and ultrasound scanners and confocal or ordinary microscopy. Measurements, display techniques and user interaction are important items for the medically oriented users. The acquisition is mostly performed by a scanning device, which can introduce the problem of distortions induced by movement on top of the distortions of lenses. Exact reconstruction requires mostly operations in the frequency domain. A problem in 3D image processing is the non-uniform resolution in x, y and z direction of the image, typically 256 x 256 x 64 volume elements or voxels. This urges the need for either resampling the image or requires the modification of existing routines to the non-uniform resolution. An example of a processing sequence in the field of cytometry would be the reconstruction of periodic signals based on FFT & IFFT (Young 1988 et al., 1989) to compensate for the acquisition distortions, followed by thresholding, three-dimensional morphologic operations and three-dimensional labelling of objects to find cell nuclei. This sequence is repeated to find centromeres in interphase cell nuclei, followed by a three dimensional distance transform (a modified approach of: Danielsson (1980), Yamada (1984)) to get the distance of the centromeres to the borders of the nuclei. The then performed measurements are the exact calculation of the center-of-mass, the surface area, the volume, texture, shape and total intensity of each object (The processing sequence of the ANA-3D benchmark for 3D image analysis workstations: Young 1990).

2.2 Low-level image processing operations.

In this section first a classification of low-level image processing operations is given, followed by some examples from each class. The classification and description method are extended versions of the classification and description method of (Komen 1990a).

2.2.1 Classification of low level operations.

Given: A two dimensional source image X, a destination image Y, a value (scalar) or value-array V, the image origin O, a structuring element or neighbourhood S and *p*, *q* the position vectors of pixels within X, Y, or S, then the following operations are considered to be low-level image operations (see also figures 2.1 and 2.2):

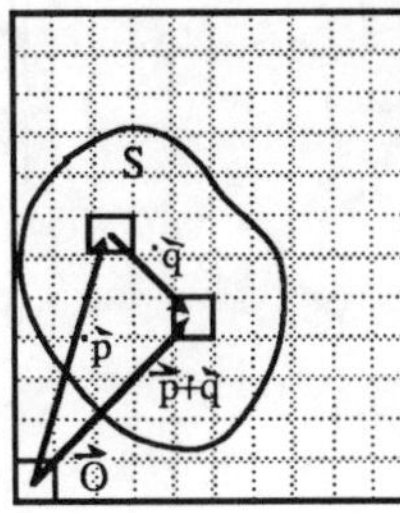

Figure 2.1 Image, pixels and structuring element within an image.

- **Point Operation (PO).** Every pixel $Y_{\vec{p}}$ in the destination image is a function of the pixel $X_{\vec{p}}$ in the source image.
- **Local Neighbourhood Operation (LNO).** Every pixel $Y_{\vec{p}}$ in the destination image is a function of the pixels $X_{\vec{p}+\vec{q}}$ in the source image. Where $\vec{q}$ are all positions within the structuring element or neighbourhood S.
- **Recursive Neighbourhood Operation (RNO).** Every pixel $Y_{\vec{p}}$ in the destination image is a function of both the pixels $X_{\vec{p}+\vec{q}}$ *(the normal pixels)* in the source image and the pixels $Y_{\vec{p}+\vec{q}}$*(the recursive pixels)* in the destination image. Where $\vec{q}$ are all positions within the structuring element or neighbourhood S. Note that the pixel on position $\vec{q} = 0$ is called the *central pixel*. The central pixel mostly does not belong to the set of recursive pixels $Y_{\vec{p}+\vec{q}}$.
- **Object Operation (OO).** Every pixel $Y_{\vec{p}}$ in the destination image is a function of the pixels $X_{\vec{p}+\vec{q}}$ in the source image. Where $\vec{q}$ are all positions within the object.
- **Global Operation (GlO).** Every pixel $Y_{\vec{p}}$ in the destination image is a function of the pixels $X_{\vec{p}+\vec{q}}$ in the source image and possibly of the positions $\vec{p}$ and / or $\vec{q}$ themselves. Where $\vec{q}$ belongs to the complete image X
- **Geometric Operation (GeO).** Every pixel $Y_{\vec{p}}$ in the destination image is a function of the pixels $X_{g(\vec{p})+\vec{q}}$ in the source image and possibly of the positions $\vec{p}$ and / or $\vec{q}$ themselves. Where $\vec{q}$ belongs to the structuring element S, which may be depending on $Y_{\vec{p}}$ and g is a function on the source coordinate $\vec{p}$.
- **Statistical Operation (SO).** The value or value vector V is a function of all pixels $X_{\vec{p}}$ in the source image. Where $\vec{p}$ belongs to the complete image.

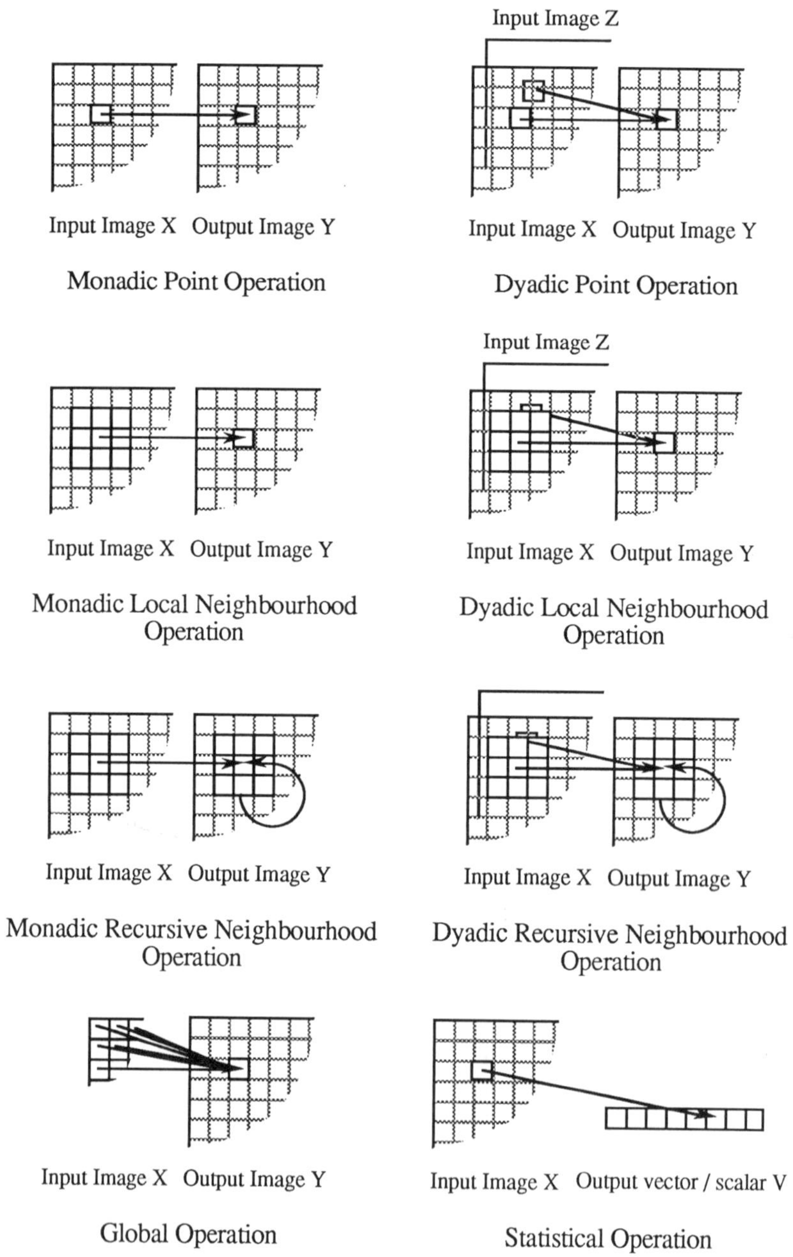

Figure 2.2 The classes of low-level image processing operations

At this point, a slight distinction can be made between the terms structuring element and neighbourhood. This distinction is, however, informal. See also section 4.1. Whereas with the name structuring element (see also: Serra 1982) no standard geometric boundary is associated, but merely a set of points able to restructure (or filter) another set, with the term neighbourhood or filter kernel is usually more associated a fixed bounded section around a central pixel of an image. Usually the neighbourhood is square, has an odd size and is rather small. Common sizes are 3 x 3, 5 x 5 to 21 x 21 pixels. As such, a neighbourhood can also be considered as an implementation of a structuring element for a digitized image.

The classification of low-level operations can be extended in a few ways.

- Some operations like the point operation, the local neighbourhood operation and the recursive neighbourhood operation not only have a **monadic form** but also also have a **dyadic form**. Figure 2.2. depicts that in addition to the input image X a second input image Z can be used. However, in the case of neighbourhood operations, only the neighbourhood of image X is involved in the calculation of the final result.
- The classification applies both to **grey-value images** (usually 8 bit images) as well as to **binary images** (1 bit values of the pixels). Local and recursive neighbourhood operations on binary images are usually referred to as cellular logic operations and consequently the special image processing machines with architectures based upon these operations are referred to as cellular logic image processors.
- The local neighbourhood operations (LNO or RNO) can also have an **n-pass form** and is based on a few passes of the operation over the image.
- The local neighbourhood operations (LNO or RNO) can also have an **iterative form** and is based on an iteration of the neighbourhood operation over the image until idempotence occurs.
- The local neighbourhood operations (LNO or RNO) can also have a **combined form.** The low-level operation is based on a combination of different neighbourhood operations over the image. Combinations with point operations are also possible.
- Finally, the neighbourhood operations can be separated into **morphological neighbourhood operations** (if they operate on the structure that is both present in the neighbourhood as well as locally in the image), and in operations

that can be characterized as **statistical neighbourhood operations**. In the latter the relation between pixel position and pixel value is not relevant.

2.2.2 Examples of low level operations.

In this sub-section some examples of low-level operations from each class are shown. The examples are discussed in sub-section 2.2.3, where the classification and the examples are used to formulate primitive functions for low-level operations.

- **Grey-value monadic Point Operations:**

$(\forall\vec{p} \in Y)\ Y_{\vec{p}} \leftarrow \mathrm{Ln}\,(X_{\vec{p}})$ *Logarithm*

$(\forall\vec{p} \in Y)\ Y_{\vec{p}} \leftarrow \sqrt{X_{\vec{p}}}$ *Square root*

$(\forall\vec{p} \in Y)\ Y_{\vec{p}} \leftarrow (X_{\vec{p}})^2$ *Square*

$(\forall\vec{p} \in Y)\ Y_{\vec{p}} \leftarrow (X_{\vec{p}}) + 1$ *Increment*

- **Grey-value dyadic Point Operations:**

$(\forall\vec{p} \in Y)\ Y_{\vec{p}} \leftarrow X_{\vec{p}} + Z_{\vec{p}}$ *Addition*

$(\forall\vec{p} \in Y)\ Y_{\vec{p}} \leftarrow X_{\vec{p}} \times Z_{\vec{p}}$ *Multiplication*

$(\forall\vec{p} \in Y)\ Y_{\vec{p}} \leftarrow \mathrm{Max}\,(X_{\vec{p}}, Z_{\vec{p}})$ *Maximum*

$(\forall\vec{p} \in Y)\ Y_{\vec{p}} \leftarrow X_{\vec{p}} > \mathrm{Thr}$ *Threshold*

(with: Thr = the threshold value)

- **Grey-value morphological Local Neighbourhood Operations:**

(with $C_{\vec{q}}$ being the elements of S; the filter coefficients)

$(\forall\vec{p} \in Y)\ Y_{\vec{p}} \leftarrow \sum_{(\forall\vec{q} \in S)} (X_{\vec{p}+\vec{q}} \times C_{\vec{q}})$ *General Convolution*

Used in a combined form with point operations for e.g. the *SOBEL edge detector*:

$(\forall\vec{p} \in Y)\ Y^1_{\vec{p}} \leftarrow \sum_{(\forall\vec{q} \in S)} (X_{\vec{p}+\vec{q}} \times C^1_{\vec{q}})$ with: $C^1 = \begin{matrix} -1 & 0 & -1 \\ -2 & 0 & -2 \\ -1 & 0 & -1 \end{matrix}$ pass 1

$$(\forall \vec{p} \in Y)\ Y^2_{\vec{p}} \leftarrow \sum_{(\forall \vec{q} \in S)} (X_{\vec{p}+\vec{q}} \times C^2_{\vec{q}}) \quad \text{with: } C^2 = \begin{matrix} -1 & -2 & -1 \\ 0 & 0 & 0 \\ -1 & -2 & -1 \end{matrix} \quad \text{pass 2}$$

$$\text{and } (\forall \vec{p} \in Y)\ Y_{\vec{p}} \leftarrow \sqrt{\left(Y^1_{\vec{p}}\right)^2 + \left(Y^2_{\vec{p}}\right)^2} < \text{Thr} \quad \text{pass 3}$$

- **Grey-value statistical Local Neighbourhood Operations:**
  (with n being the number of elements of S)

$$(\forall \vec{p} \in Y)\ Y_{\vec{p}} \leftarrow \text{Floor}\left(\frac{1}{n} \times \sum_{(\forall \vec{q} \in S)} (X_{\vec{p}+\vec{q}})\right) \quad \textit{Uniform filter}$$

$$(\forall \vec{p} \in Y)\ Y_{\vec{p}} \leftarrow \text{Max}_{(\forall \vec{q} \in S)} (X_{\vec{p}+\vec{q}}) \quad \textit{Local Maximum filter}$$

$$(\forall \vec{p} \in Y)\ Y_{\vec{p}} \leftarrow \left(\text{Sort-up}_{(\forall \vec{q} \in S)} (X_{\vec{p}+\vec{q}})\right)\left[\frac{n+1}{2}\right] \quad \textit{Median Filter}$$

Used in a combined form with point operations for e.g. the *Dynamic Threshold:* (see e.g. Verbeek et al. 1988)

$$(\forall \vec{p} \in Y)\ Y^1_{\vec{p}} \leftarrow \text{Max}_{(\forall \vec{q} \in S)} (X_{\vec{p}+\vec{q}}) \quad \text{pass 1}$$

$$(\forall \vec{p} \in Y)\ Y^2_{\vec{p}} \leftarrow \text{Min}_{(\forall \vec{q} \in S)} (X_{\vec{p}+\vec{q}}) \quad \text{pass 2}$$

$$(\forall \vec{p} \in Y)\ Y_{\vec{p}} \leftarrow \frac{1}{2}(Y^1_{\vec{p}} + Y^2_{\vec{p}}) \quad \text{pass 3}$$

- **Grey-value morphological Recursive Neighbourhood Operations:**
  Distance transform, (see e.g. Borgefors 1986)

$$(\forall \vec{p} \in Y)\ Y^1_{\vec{p}} \leftarrow \text{Min}_{(\forall \vec{q} \in S)} (X_{\vec{p}},\ Y^1_{\vec{p}+\vec{q}} + C^1_{\vec{q}}) \quad \text{pass 1} \quad \text{with: } C^1 = \begin{matrix} 7 & 5 & 7 \\ 5 & 0 & 0 \\ 0 & 0 & 0 \end{matrix}$$

$$(\forall \vec{p} \in Y)\ Y^2_{\vec{p}} \leftarrow \text{Min}_{(\forall \vec{q} \in S)} (Y^1_{\vec{p}},\ Y^2_{\vec{p}+\vec{q}} + C^2_{\vec{q}}) \quad \text{pass 2} \quad \text{with: } C^2 = \begin{matrix} 0 & 0 & 0 \\ 0 & 0 & 5 \\ 7 & 5 & 7 \end{matrix}$$

- **Grey-value statistical Recursive Neighbourhood Operations:**
  Recursive median, if applied iteratively until idempotence: *median root*. (See e.g. Komen 1990a)

$$(\forall \vec{p} \in Y)\ Y_{\vec{p}} \leftarrow \left(\text{Sort-up}_{(\forall \vec{q} \in S)} (Y_{\vec{p}+\vec{q}})\right)\left[\frac{n+1}{2}\right]$$

- **Grey-value Object Operations:** p.m.

- **Grey-value Global Operations:**
  Two dimensional FFT (Cooley and Tukey 1965)

$$(\forall \vec{p} \in Y)\ Y^{1}_{\vec{p}} \leftarrow \sum_{(\forall \vec{q} \in X)} (X_{\vec{q}} \times C^{1}_{\vec{p},\vec{q}}) \qquad \text{pass 1 over the rows}$$

$$(\forall \vec{p} \in Y)\ Y^{2}_{\vec{p}} \leftarrow \sum_{(\forall \vec{q} \in Y^{1})} (Y^{1}_{\vec{q}} \times C^{2}_{\vec{p},\vec{q}}) \qquad \text{pass 2 over the columns}$$

with: $C^1 = \boxed{e^{-j(2\pi/N)p_r q_r}}$, $C^2 = \boxed{e^{-j(2\pi/N)p_c q_c}}$ and p_r and q_r the row coordinates of $\vec{p}$ and $\vec{q}$ and p_c and q_c the column coordinates of $\vec{p}$ and $\vec{q}$. N x N is the size of the image. All arithmetic operations apply on complex numbers ($j^2 = -1$).

- **Grey-value Geometric Operations:**

$$(\forall \vec{p} \in X)\ \ Y_{(\text{round}(\vec{p} \cdot \vec{r} + \vec{t}))} \leftarrow X_{\vec{p}} \qquad \textit{Affine transformation}$$

See also (Gerbrands 1988) for applications of more complex geometric transformations, i.e. polar transformations.

- **Grey-value Statistical Operations:**

$$(\forall \vec{p} \in X)\ \ V[X_{\vec{p}}] \leftarrow V[X_{\vec{p}}] + 1 \qquad \textit{Histogram calculation}$$

- **Binary Monadic Point Operations:**

$(\forall \vec{p} \in Y)\ Y_{\vec{p}} \leftarrow \text{Not}\ (X_{\vec{p}})$ *Logic Inverse*

- **Binary Dyadic Point Operations:**

$(\forall \vec{p} \in Y)\ Y_{\vec{p}} \leftarrow (X_{\vec{p}}\ \text{And}\ Z_{\vec{p}})$ *Logic And*

- **Binary monadic morphological Local Neighbourhood Operations:** (with $C_{\vec{q}}$ being the elements of S; the hit-or-miss filter coefficients and $\cong$ the inexact match, a combined Boolean operation. See also chapter 4.1 for a more extended explanation of the hit-or-miss transform)

$(\forall \vec{p} \in Y)\ Y_{\vec{p}} \leftarrow \bigwedge_{(\forall \vec{q} \in S)} (X_{\vec{p}+\vec{q}} \cong C_{\vec{q}})$ *General Hit-or-Miss Transform*

and used with: $C = \begin{matrix} 1 & 1 & 1 \\ 1 & 1 & 1 \\ 1 & 1 & 1 \end{matrix}$ *Erosion of 4 connected contours*

- **Binary monadic statistical Local Neighbourhood Operations:**

$(\forall \vec{p} \in Y)\ Y_{\vec{p}} \leftarrow \text{Thr} > \sum_{(\forall \vec{q} \in S)} X_{\vec{p}+\vec{q}}$ *Majority Vote or Binary Rank Filter*

- **Binary monadic morphological Recursive Neighbourhood Operations:**

$(\forall \vec{p} \in Y)\ Y_{\vec{p}} \leftarrow Z_{\vec{p}}\ \text{And}\ (\text{Not} \bigwedge_{(\forall \vec{q} \in S)} (Y_{\vec{p}+\vec{q}} \cong C_{\vec{q}}))$

used with: $C = \begin{matrix} 0 & 0 & 0 \\ 0 & 0 & 0 \\ 0 & 0 & 0 \end{matrix}$ *Propagation*

- **Binary Global Operations:**

$V \leftarrow \bigvee_{(\forall \vec{p} \in X)} X_{\vec{p}}$ *Global Or*

$$V \leftarrow \sum_{(\forall \vec{p} \in X)} X_{\vec{p}}$$ *Pixel count*

2.2.3 Discussion on the requirements for low-level operations.

One of the first evident observations is that a mechanism must be present to visit all pixels in the input image $(\forall \vec{p} \in X)$, the output image $(\forall \vec{p} \in Y)$ and the neighbourhood $(\forall \vec{q} \in S)$. Note that is it excluded that input and output images have a different size or that there is a restriction imposed on the size of the images.

For Point Operations and statistical Local Neighbourhood Operations the sequence in which the pixels of the image are updated does not matter. For morphologic Local Neighbourhood Operations the result of the operation may be shifted one pixel in either direction due to the update technique used. (See also chapter 4, on the thinning subject). For Recursive Neighbourhood Operations the update technique may influence the whole resulting image, however, this is data dependent. (For an extended information of this subject see Komen 1990a.)

Most common update techniques are:

- Raster update: for y := starty to endy do
  { for x := startx to endx do *visit pixel* }
- Meander update: for y := starty to endy do
  { if y = even do {for x = startx to endx do *visit pixel* }
  else do {for x = endx downto startx do *visit pixel* } }
- Parallel update: All pixels of the image are updated in parallel.
- Line-wise update: All pixels of one image line are updated in parallel, the lines serial.

The first two techniques are sequential update techniques usable in software or in data flow pipeline machines. The last two techniques are used in data parallel machines, in the Square Processor Array and the Linear Processor Array respectively. (See also chapter 3.)

Generally the update method of the neighbourhood itself is not important. The maximum size of the neighbourhood is equal to the input image size.

A neighbourhood may be scarcely occupied, e.g. only the pixels on a certain radius around the central pixel are involved in the operation (e.g. see Verbeek et al. 1988).

For recursive neighbourhood operations, the available neighbourhood depends on the image update technique that is used, as only pixels in the image that have been updated so far can be used in the recursive neighbourhood. Note that the size and shape of the available recursive neighbourhood may influence the features and possibilities of the Recursive Neighbourhood Operations.

Many Global and Object Operations can be written as Local or Recursive Neighbourhood Operations. (See for extended information on this subject: Komen 1990a). As an example of an RNO that only affects the objects in the image, the distance transform can be taken (Borgefors 1986).
From this point on we will classify any Global Operation or Object Operation that cannot be written as a Point Operation, a Local or Recursive Neighbourhood Operation, a Geometrical Operation, a Statistical Operation or a combination of these as an intermediate-level algorithm.

As the neighbourhoods may have an arbitrary size, albeit not always fully occupied, a solution must be found to handle the neighbourhood processing at the image boundaries. Common procedures for neighbourhood points that fall outside the image boundary $\vec{p} + \vec{q} \notin X$ are:

- Replace those pixels by a constant image boundary or image edge value:

  $$X_{\vec{p}+\vec{p}} \leftarrow \text{edge\_value}$$

- Replace those pixels by the value of the pixels mirrored around the image boundary.

  $$X_{\vec{p}+\vec{q}} \leftarrow X_{(\vec{n}-\vec{p}-\vec{q}) \bmod \vec{n}}$$

- Replace those pixels by the value of the pixels wrapped around the image.

  $$X_{\vec{p}+\vec{q}} \leftarrow X_{(\vec{n}+\vec{p}+\vec{q}) \bmod \vec{n}}$$

- Replace those pixels by the value of the pixels propagated from the boundary.

  $$\forall_i \left(X_{p_i+q_i} \leftarrow X_{\min(n_i,\, \max(O,\, (p_i+q_i)))} \right)$$

With $\vec{O}$ being the origin of the image, $\vec{n}$ the vector pointing to the maximum image size and i the index of the vector components.

Consequently, for the neighbourhood **address processing**, operations like adding, subtracting, modulo, minimum and maximum are needed.

For the processing of the pixels themselves, operations like adding, subtracting, increment, decrement, multiplication, minimum, maximum, round, floor, ceil, modulo and various look-up table actions for logarithm, sine, square root, e^x, etc. are necessary.
Statistical neighbourhood operations are using accumulative arithmetic (e.g. addition), but also operations such as less than, greater than, equal to, not equal to and all possible Boolean operations between 2 variables.
For median filtering or more generally rank or percentile filtering a sorting mechanism would be profitable. However, median filtering can also be implemented as a sequence of maximum and minimum operations. See section 5.5.2.
One of the most important subjects for histogram calculations and geometric operations is array indexing or indirect addressing. For geometrical operations, where extensive address calculations take place, it would be profitable if pixel addresses could be treated as normal data. In that case swapping or merging of data and addresses would be a necessity.
From the low-level operations that are combinations of other low-level operations, n-pass or iterative, it can be deduced that in some cases, operations can be applied in parallel, after which the result can be joined in a next operation and in other cases the operations have to be performed sequentially. In the first case, an operation parallel approach can be used, in the latter, pipelining of operations can be used. For more extensive information on the sorts of parallelism and pipelining we refer to Chapter 5.
For both n-pass and iterative low-level operations, iteration counting is necessary. For iterative operations idempotence detection is necessary; the output images of the last two iterations have to match exactly. A global-or on the difference image can be used here.
More complicated global operations like the FFT, use for their implementation a sequence of operations based on arithmetic on complex numbers and the swapping of data streams, i.e. butterfly operations. (See for more extensive information on this subject e.g. Lindskog 1988). These operations require an extensive possibility to manipulate with pixels on distances larger than their nearest neighbour distance.

2.3 Description of some intermediate level operations.

In this section, four intermediate-level image processing algorithms are described, followed by a discussion on the requirements that intermediate-level algorithms impose on a computer architecture for low and intermediate level image processing.

2.3.1 Recognition based on the matching of line segment graphs.

In many industrial applications the objects under investigation have firm boundaries and consequently recognition based on the edges of objects is very profitable. The method described here is based on edge detection with low-level image processing tools followed by the construction of edge segment lists.

The segmentation (Buurman and Duin 1989) was based on a: 17 x 17 binomial filter with standard deviation 2, a 5 x 5 Dynamic Contour filter based on a non-linear Laplace zero crossing method (Verbeek et al.1988) and skeletonization based on the distance transform (Verwer 1988). The graph building procedure (see Groen et al. 1985) starts with scanning the image until a non-end or non-branch pixel of a skeleton is encountered. The curve is followed in both directions and its chain code is stored until a branch or end-pixel of the skeleton is detected. Next, the chain coded curves are approximated with a number of straight line segments. Each line segment has two end-points marked as vertices. With these segments a graph is set up consisting of a vertex list and a segment list. Both lists are linked by pointers. In a (global) recovery stage connections between segments are established and redundancies are removed. The graph is updated in the sense that vertices within a certain distance are merged and segments with a small difference in angle are connected, also new branch-points may be found by connecting vertices and segments on close distance. Finally, the graph may be pruned with some criterion such as lengths of segments, to reduce the graph for the next processing step. A possible next step would be pattern recognition based on the comparison of the graph of a test image and the graphs of images from a database. Graph comparison can then be based on comparing each segment of the test image with each segment of a database image.

2.3.2 The A* algorithm applied on path finding.

The goal directed graph search algorithm A* (see e.g. Pearl 1984) can be applied on various problems in the image domain as well. Below an example is given of an application in a high dimensional equidistant sampled grid or high dimensional image, used as the configuration space of a robot for robot path planning.
The problem definition of robot path planning is (Lozano-Pérez 1987): "Given a moving object, (A), a set of stationary objects, (B_i), at known positions, a legal initial position of A, s, and a legal goal position of A, g, find a sequence of motions that take A from s to g without colliding with any of the B_i." Note that this is static path planning, the definition does not include planning of servo actions that minimize the travel time along or close to a predefined path. Collision avoidance is an extension of the path planning problem. An uncertain environment with moving obstacles is assumed then. A quantized configuration space (Lozano-Pérez 1983) is a space in which each point represents a unique position of a robot in the real world. Each adjacent pair of points in configuration space receives a label, prohibited or permitted, dependent on whether or not a movement of the robot between those points would result in a collision. Adjacent points in configuration space are points which coordinates do not differ more than 1 in each dimension. The cost to travel from a point to its nearest neighbor point on the coordinate grid, is called the transition cost d.
A standard heuristic search method (A*) can be used in the configuration space. The nodes of the graph are points in an N-dimensional configuration space. All nodes are locally connected, except for obstacle points, which are not connected at all, as is depicted in figure 2.3. The metric in the configuration space can be chosen to approximate the euclidean distance, e.g. in a 2D case by chosing the cost d = 5 for horizontal and vertical transitions and d = 7 for diagonal transitions (Dekker et al. 1987, Verwer 1990).

Heuristics information is incorporated by an evaluation function *f*, which is the sum of two functions *g* and *h*. In graph terminology, *g(n)* represents the costs from the start node *s* to node *n* and *h(n)* represents an estimate of the costs from node *n* to the goal node *g*. A* is guaranteed to find the cheapest path if *h(n)*, the heuristic function, is chosen optimistically.E.g take for *h(n)* the 'as the crow flies' euclidian distance to the goal, i.e. ignoring the objects in the image. If *h(n)* is assigned the value zero, the A*-algorithm degenerates into a uniform cost algorithm. In that case, waves of equal cost are propagated from the start node in all directions, until the goal node is reached. Let d be defined as the maximum still optimistic value of the costs from node n to the goal node g. If $h(n)$ is increased to d the algorithm first expands nodes in the direction of the goal node, and later, when obstacles are encountered, nodes in other directions. The applicability of heuristic information therefore decreased with the complexity of the search space.

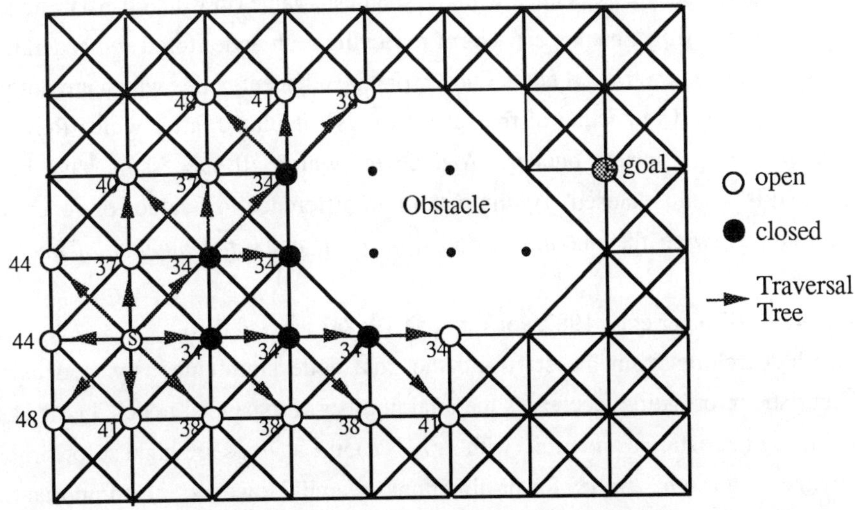

Figure 2.3. The graph of a 2D path finding process after a few node expansions. The numbers attached to the nodes are evaluation values as computed by an A* algorithm. The transition cost (d) of the diagonal direction is 7, the horizontal and vertical costs 5.

The A*-algorithm can be described in 4 steps, see figure 2.3:

1) Mark the start node s 'open' and calculate an evaluation function $f(s)$.
2) Select the open node n with the lowest f value. If a tie occurs, resolve in favour of a goal node.
3) If n is a goal node, mark n 'closed' and terminate the algorithm.
4) Otherwise, mark node n 'closed', calculate f for each successive node of n and mark 'open' each successor node not already marked 'closed'. Re-mark as 'open' any closed node n_i which is a successor of n for which $f(n_i)$ is smaller now than it was when n_i was marked 'closed'. Goto step 2.

The most calculation intensive part of the A*-algorithm is the selection of the node with the smallest *f* value. A sorting algorithm is required here. Bucket sort is the most efficient sorting algorithm, linear in the number of nodes to be sorted (Aho et al. 1974). It can be used with bounded, integer valued transition costs. The 'open' nodes are stored in buckets. Each bucket has an *f* value associated with it; nodes with equal *f* values are stored in the same bucket. The buckets are processed one by one, starting with the bucket associated with the smallest *f* value (in which the start node *s* is stored). Processing a bucket consists of retrieving each node stored and generating successors for each retrieved node. Successors are adjacent nodes which are not yet labelled 'closed'. The *f* value of the successors is calculated, after which they are stored in the appropriate buckets. A node for which all successors have been generated is labelled 'closed'. As the maximum difference in *f* values of generated successors is twice the maximum transition cost, *a limited number of buckets suffices.*

Simulations (Dekker et al. 1987, Jonker et al. 1988a) showed that:

- Both searching from the start and the goal node simultaneously, and using heuristic information, decreases the total number of expanded nodes. In the case of using heuristic information (with $h(n) = d$), the savings are highest for simple images. The same applies for bi-directional search. However, most important is that the mechanisms impede each other. Employing a bi-directional search if heuristic information is used degrades the performance, relative to the case where only heuristic information is used.
- When using directional information, a priori pruning of the search graph is possible. The gain is a reduction in the mean number of successors. Normally, nodes are opened many times before they receive the label 'closed'. If directional information is used this inefficiency diminishes. All nodes adjacent to a generating node that can be reached via a cheaper path from the node which generated the generating node (see figure 2.4) are not considered as successors. Of course, one has to check if the cheaper path is not prohibited.

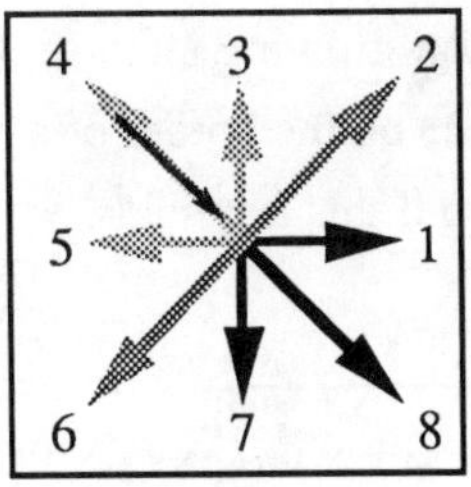

Figure 2.4 Directional information can eliminate successors. Suppose the central node was generated from neighbour 4. Neighbour 4 does not need to be considered as successor, neighbours 2,3,5 and 6 only if the transitions from neighbour 4 to neighbours 3 and 5 or to neighbours 2 and 6 via neighbour 3 and 5 are prohibited.

2.3.3 Signed Euclidian distance transform.

In automated microscopy the acquired three dimensional images are usually not uniformly sampled. Now it becomes important to perform operations on the position vectors of the volume elements (voxels) rather than on the values of the voxels themselves. In the signed euclidian distance transform as introduced by (Danielson 1980) and amended by (Yamada 1984) the distance propagation is performed on the position vectors from an object pixel to its nearest background pixel. In the update of the position vectors and calculation of the final distance image from the position vectors, the uneven sampling density can be taken into account.

2.3.4 Region growing applied on segmentation of range images.

The A* algorithm applied on images and the distance transform can both be denoted as wave front algorithms. Quite similar but still a different wave front problem is the region growing problem for 2.5D images. (For a general introduction in region growing see Ballard and Brown 1982.) Applied on range data segmentation, first and second order surfaces are fitted onto a 2.5D 'landscape' (Besl and Jain 1986, 1988). See figure 2.5.

In the implementation of (Schmidt 1989) the range image is first filtered to suppress high frequencies (a low-level operation). Then a binary occlusion image is set-up. This image contains the points in the image that have invalid data due to the fact that these points were occluded. A seed image is then generated, and from each seed in this image a homogeneous region, a first order surface, should be grown. The seed generation is again a low-level filter algorithm that picks out all possible candidates for flat regions. Then the region growing starts from these seeds and a number of wave-fronts occur. All the points on the border of a region are tested with a growing criterion and added to the region if this test holds.

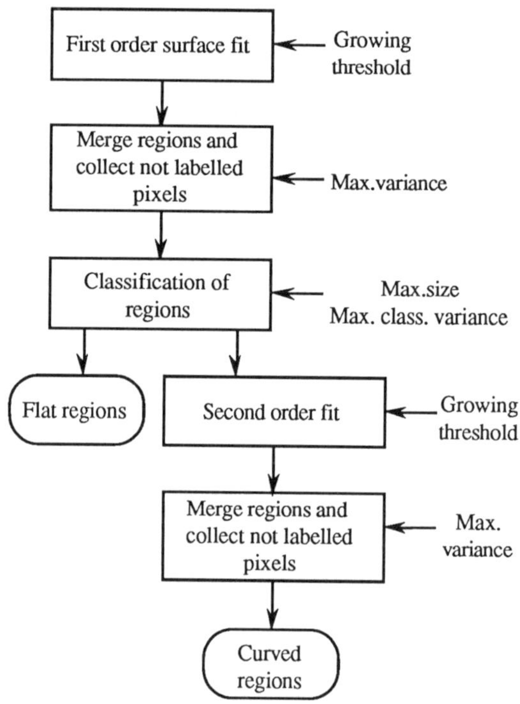

Figure 2.5. Region growing used in 2.5D image processing (Schmidt 1989).

Equation (2.1) gives a general equation of a surface in $\mathbb{R}_3$, (2.2) gives the growing criterion for planar surfaces in a 2.5D image and (2.3) the growing criterion for quadratic surfaces.

$$x_3 = \sum_{m=0}^{M} \sum_{i=0}^{m} a_{m,i} \cdot x_1^{i} \cdot x_2^{m-i} \qquad (2.1)$$

$$(a_{00} + a_{10}x_1 + a_{01}x_2 - x_3) \leq gr\_thr_1 \qquad (2.2)$$

$$(a_{00} + a_{10}x_1 + a_{01}x_2 + a_{11}x_1x_2 + a_{20}x_1^2 + a_{02}x_2^2 - x_3) \leq gr\_thr_2 \qquad (2.3)$$

With x_1 and x_2 the coordinates of a point in the 2.5D image, and x_3 the value at those coordinates. The growing thresholds are parameters to be chosen.
During the testing of the border points with the growing criterion, the surface parameters $a_{m,i}$ are not updated and therefore all border points of a region are tested with the same criterion. If the whole border has been processed, a set of equations is solved to update the surface parameters $a_{m,i}$. These equations form the solution of the parameter estimation performed by minimizing the sum (see 2.4 for the first order fit) in least squared sense.

$$\text{Error} = \sum_{\text{region}} (a_{00} + a_{10}x_1 + a_{01}x_2 - x_3)^2 \tag{2.4}$$

In order to speed up calculations after the front of a region is totally updated (minimizing (2.4)), partial sums and sums of squares can be maintained and updated after each allocation of a point to a region. The growing procedure for a region continues until no more points can be added to that region.
Figure 2.5 depicts the whole region growing procedure. Note that after each fit, global post-processing takes place, in the sense that two or more neighbouring regions are merged if their error is below a certain threshold (this can occur due to poor seed generation). Also, not classified single pixels or pairs of pixels can be merged with the neighbouring region that comes closest to the growing threshold (2.2 or 2.3), and between the first and second order fit, a classification of regions takes place. All small regions and regions with a large variance of fit are classified as candidates for the higher order fit.
The outcome of this region growing procedure is a labelled image. Each label indicates a surface; all points of the same surface have an identical label. Associated with a label are the parameters of its surface. Edges in the 2.5D image can be obtained by geometrically intersecting the surfaces that are found in the region growing procedure.

2.3.5 Requirements for intermediate-level operations.

One of the most pertinent subjects in the previous examples is the transformation from the image data-structure to a list or tree like data structure that is used in combination with the normal image array structure. The lists are filled by either a raster scan through the image, for example to find the border points, skeleton endpoints, etc. or by a guided walk along skeleton branches. In the latter case, a mechanism must be present to transform a hit on a branch to a set of freeman codes of the pixels that are set in its neighbourhood. This code can be used to retrieve the addresses of those neighbours. For graph building a mechanism to set-up and maintain queues must be present, while within the queues, storage of data as well as addresses (pointers) should be allowed. With these pointers other queues can be addressed.

A bucket queue is a heap of normal queues on which the base address of each queue or bucket is used to access it. Usually this base address is meaningful data, like the distance of a pixel to the border of an object and the bucket is for instance filled with the addresses of all pixels having that distance. A mechanism must be provided to retrieve data from the bucket queue in an orderly way, for example: smallest distance first.

Bucket queues can be used to store graphs and can be used to sort and store the pixels of a wave front. As such, they can be used to process only the pixels that are likely to change in the image (Groen et al. 1988). Distance transforms, cellular logic operations and search algorithms can be based on the bucket queues. The processing now can be performed by walking through a queue and performing operations on the entries of that queue and (simultaneously) on other queues, transforming the queue into another.

Another relevant subject from the examples above, is that the data on which is operated is not solely image data, but in addition to that parameters, such as the surface parameters in the region growing example, or the position vectors in the non-uniformly sampled grid of the signed euclidean distance transform example. In both cases a word length of 32 bits, 64 bits or even floating point is necessary. In many situations, scaling to integer with a large word length can be profitable, however, it cannot always be avoided, e.g with anisotropic sampling densities in automated microscopy. In the latter case, where in the final processing stage in the geometrical domain measurements are performed, floating point calculations are a necessity. This leads to the conclusion that in many intermediate-level algorithms floating point must be at hand, albeit with a possible penalty for its use.

2.4 Requirements for image processor architectures.

A final note on high-level image processing algorithms. These can usually be seen as sub-tasks of applications such as industrial inspection systems, robot vision systems and interactive biomedical workstations. In many situations the high-level algorithms involve the construction of intelligent controllers operating on a toolbox of image processing routines and services. General purpose for the low-level routines, more or less dedicated for intermediate level routines. The decision which routine to use and when, and which parameters to choose is made by the controller, that can be based on an expert system using expert knowledge (Ozaki 1988, Cheng 1990). Such a controller is able to start up competing procedures, weighting the results, or combining evidence. This usually means that the execution of whole sub-tasks can be performed in parallel, which suggests the use of coarse grain parallelism for this level.

3D image processing involves the processing of many more elements than the usual 2D or 2.5D images. In many applications, however, even the 2D data involved becomes huge.
Undoubtedly the parallelism that can be exploited in the low-level routines is characterized as fine grain parallelism. The parallelism that can be exploited in intermediate-level routines rather depends on the algorithm. Recent case studies point out that services to enable the parallel processing of queue like structures would be beneficial.

2.5D image processing involves triangulation and may use geometric operations. Automated microscopy and many other applications in which measurements are the final aim, involves exact reconstruction of the original image and, calculating and maintaining the exact geometrical positions in the image. Floating point support for these tasks must be at hand.

Most image processing sub-tasks demand a mixture of low-level and intermediate level image processing operations and purely low-level tasks are at the present exceptional. Consequently, novel computer architectures for image processing should smoothly support both low-level as well as intermediate-level image processing. That the interfacing with host systems, image acquisition and image display units should be seamless is indisputable.

Furthermore, whilst workstations become faster annually and clever sequential programming remains a skill of many, the challenge of programming an odd parallel machine, while not having the freedom and services a more or less standard sequential architecture offers is often minimal. If the novel machine is not programmable in a common high level language, it has little chance tot get adopted. Random access to each single pixel and the possibility to set up any data-structure is essential. The competition of a sequential workstation is high.

3 Image processing architectures.

After the review of possible image processing tasks and algorithms of chapter 2, in this chapter a review of existing machines is presented. The aim of this chapter is to show their architectural solutions with an emphasis on possibilities for low-level and intermediate level image processing.
For general literature on (parallel) computer architectures see also: (Hennessy and Patterson 1990, Tanenbaum 1990, van de Goor 1989, Hwang and Briggs 1989).
For image processing architectures, see also: (Fountain 1987, Komen 1990a)

Section 3.1 presents the used classification scheme according to which the machines are presented.
Section 3.2 describes the abilities of uniprocessors, vector processors and Digital Signal Processors.
Section 3.3 describes the abilities of MIMD systems.
Section 3.4 describes the abilities of SIMD systems.
Section 3.5 describes the abilities of Pipeline systems.

The chapter ends in section 3.6 with an evaluation of the discussed systems.

3.1 Classifications of architectures.

In the previous years, many schemes have been set up to classify computer architectures. A brief survey of these classification schemes has been made by (Komen 1990a), leading to a detailed classification system for low-level image processing architectures, incorporating many forms of parallelism.
As the purpose of this chapter is to give a summary of the types of architectures, the used classification can be less detailed.

One of the oldest classification schemes for computer architectures is the streams model (Flynn 1972). The model separates a computer into Processing Element (PE) and Control Unit (CU) and segregates them into four classes: Single Instruction-stream Single Data Stream (SISD), Multiple Instruction-stream Multiple Data Stream (MIMD), Single Instruction-stream Multiple Data Stream (SIMD) and Multiple Instruction-stream Single Data Stream (MISD).
Although this model is augmented and amended many times, its simplicity keeps it alive; especially the terms MIMD and SIMD are widely used. The terms SISD and

MISD are continuously mentioned in literature but never practically used. SISD can be replaced by a uniprocessor system and although Pipeline systems are occasionally referred to as MISD, the term MISD will be neglected in this thesis.

Architectures can be coarsely sectioned, based on:

1) The task granularity.
2) Data parallel processing c.q. data flow processing.
3) Single or multi resolution processing.

1) Task granularity:
Emphasizing the tasks to be performed, the granularity of the main computing task usually defines the amount and sort of parallelism that can be used. With the task-classes ordered from coarse grain parallel to fine grain parallel, the architecture-classes can be ordered from loosely coupled to tightly coupled systems. The coupling degree expresses, to what extent information has to be exchanged. This information can be data as well as control (program) synchronization. In loosely coupled systems, the exchange rate between system parts in terms of amount of data, data transfer speed and transfer occurrence is generally low. In tightly coupled systems, the exchange rate between system parts is generally high.

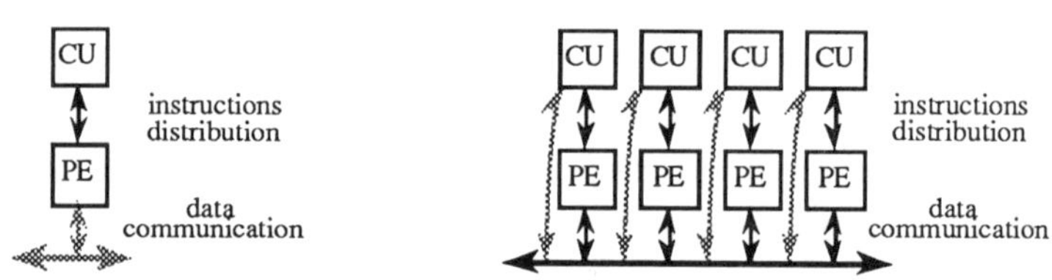

a) Uniprocessor b) MIMD system

Figure 3.1 Loosely coupled systems.

Figure 3.1 shows the outlines of loosely coupled systems for coarse grain parallel tasks, the uni/multi processor and the MIMD system.
Most common computers systems are uniprocessors, on which the operating system takes care of the coarse parallelism necessary for multiuser and multitasking operation. Multiprocessor systems (systems with one or more identical processors on a common bus) fall also into this class, as the distribution of tasks over the processors is left to the operating system and not to the user. Uniprocessors equipped

with special processors (e.g. vector processors) to boost the performance fall into this class for the same reason.
The intended use of an MIMD system is that one single coarse grain task can be split up over various processors, whereas the exchange rate is so low that the data communication network forms no severe bottleneck for the performance of this task. Characteristic of MIMD systems is, that the communication is controlled by the programs of the communicating processing elements, although the details of the communication process may be automatically performed by the PEs. Note that for synchronization of PE programs (exchange of control) usually the same data network is used. Generally one PE is designated as the processor that boots the system (the primus inter parus) and normally, one PE is designated as the processor that starts and supervises the task.

2) Data parallel processing vs. data flow processing.
In data parallel processing, one instruction, program or task (depending on the granularity) runs in parallel on all data from a single data-structure. As a consequence, the instruction, program or tasks have to be fed in a sequence to the system parts.
In data flow processing, a single data element, data structure or data set (depending on the granularity) runs in a sequence through all system parts. As a consequence, the instruction, program or tasks have to be furnished in parallel to the system parts.
Usually the terms data parallel processing and data flow processing are only used in combination with fine grain tasks. In coarse grain tasks e.g. running on an MIMD machine the principle is often applied, but rarely designated as such.
Within fine grain tasks, the data parallel approach is also referred to as massively data parallel or SIMD processing. The data flow approach is usually referred to as Pipelined processing, whereas the name data flow computing usually is left to pipelined process*ing*, in which the control is distributed with the data and usually flows along with the data. (see also section 3.5.5) In this thesis we will use the designations SIMD system, Pipeline system and dataflow computing.

Figure 3.2 shows a) the SIMD system and b) the Pipeline system. The main embodiments of SIMD systems for image processing are the Square Processor Array (SPA) and the Linear Processor Array (LPA). The Square Processor Array has a two dimensional interconnection network that maps well onto the data structure of a two dimensional image. Each PE is connected to its 4 or 8 nearest neighbours. The Linear Processor Array has a less complicated data network structure and can be used in a scanning mode to process the image. Each PE is connected to its 2 nearest neighbours. (See also Chapter 5)

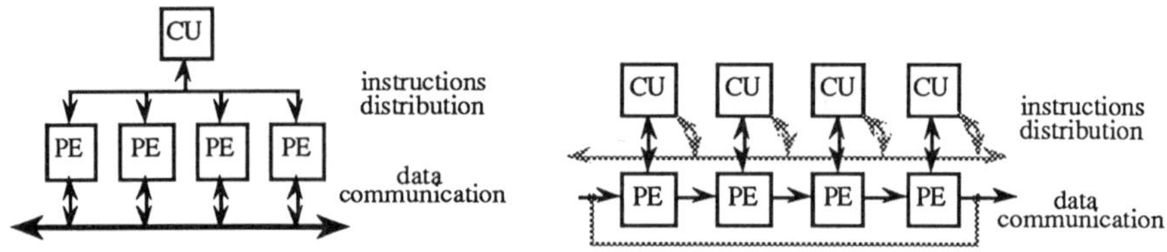

a) SIMD system b) Pipeline system

Figure 3.2 Tightly coupled systems.

The name Linear Processor Array or Linear Array is occasionally confused with a Pipeline system. One of the natural implementations of the data network of the LPA is the shift register, which makes the step to Pipeline processing short. In such cases we will use the term "LPA system in pipeline mode".
Like the SIMD system, the Pipeline system has several embodiments. However, it is important that although each control unit runs its own sequence, usually another path exist (e.g. a host computer bus) in which -normally infrequently- control information can be exchanged between the PEs. This is generally supervised, initiated and / or controlled by the host. Data dependent control can also be transported to PEs next in the line, commuted as data. Note that ring systems can considered to be Pipeline systems.
The data transfer between the PEs can be performed asynchronous, but is usually synchronous to gain speed. A special variant of a Pipeline is the systolic array as showed in figure 3.3.

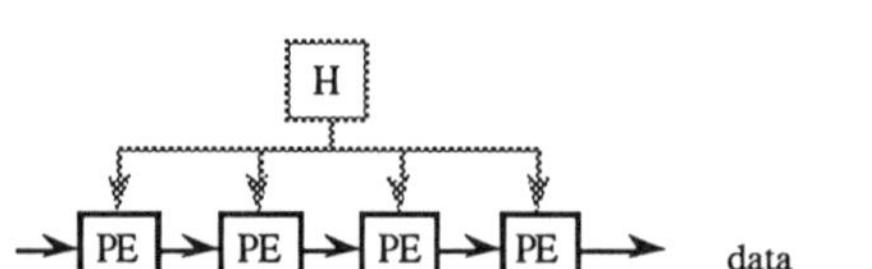

Figure 3.3 A systolic array.

A systolic array can be perceived as a synchronous pipeline system in which all PEs run the same hard wired program. We will refer to systems labelled "programmable systolic system" as Pipeline systems.
Finally, as the term Real-time system can be used for fine grain systems that process data at video rate, both SIMD systems and Pipelines can be principally used as real-time systems. For a discussion on this topic we will refer to chapter 5. The term Video Pipeline is used for synchronous pipeline systems that process at video rate.

Multi resolution processing:
In multi-resolution systems, an instruction or program is iterated over a data structure with an up or down sampling phase of the data after each iteration.
Although the operation does not change, the repeated operation usually changes the meaning of the data (apart from the reduction in the amount of data) with sometimes as a consequence that a succeeding task has a lower granularity.
This process is often referred to as Pyramidal processing, leading to the term Pyramid systems or Pyramids, for computer architectures that support this multi resolution processing. Pyramidal systems can be implemented both in an SIMD approach, for example as a stack of gradually smaller SPAs and in a Pipelined way, using interpolation and extrapolation of data streams.
The process that, due to operations, the amount of data changes and simultaneously its meaning changes to a higher abstraction level, is often reason to conceive a system that operates in this way as a pyramidal system. For example to be implemented with a system consisting of a square SIMD processor coupled to a square MIMD processor, coupled to a uniprocessor. As in most cases the purpose of processing is the rise of abstraction level of data for adequate decision making, inevitably yielding a lower granularity, we will only refer to systems explicitly performing up and down sampling within the same data structure as pyramidal or multi-resolution systems.
Consequently, examples from the following architectural classes are presented:

Loosely coupled parallel processors:

- Uni- and Multi-processors.
- MIMD systems.

Tightly coupled parallel processors.

- SIMD systems.
- Pipeline systems.

3.2 Uni- and Multi-processors.

3.2.1 Risc architectures.

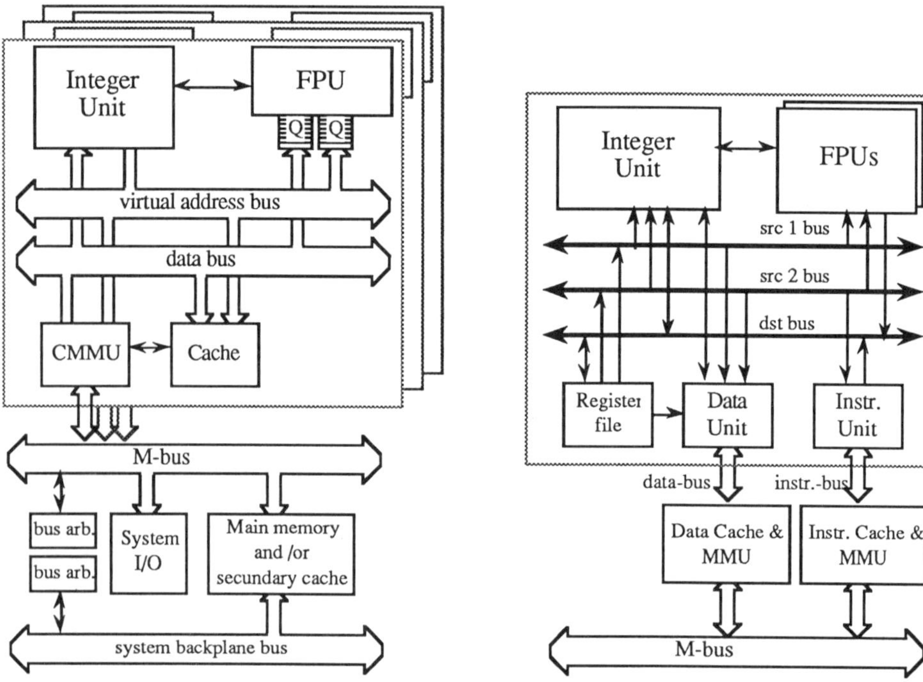

a) SPARC b) MC88x00

Figure 3.4 Uni-processor RISC architectures.

Figure 3.4a shows the SPARC architecture (SPARC 1990). Important characteristics of this architecture is the Integer Unit (CPU), Floating Point Coprocessor and large data Cache ($\simeq$ 64K), coupled onto a virtual address bus. A Cached Memory Management Unit takes care of the address translation from Virtual address to Physical address. The address translation is performed by a 4 level deep table tree walk, performed for each address not stored in its internal cache. The last 64 address translations are stored in an internal address cache to speed up address translations. The system is coupled onto the board bus (M-bus) straight to main memory or a via secondary data cache (physical address side!). For multi-processor operations more SPARC modules can be coupled onto the M-bus. In that case, a special protocol in the CMMUs takes care of the data consistency between the caches of the processors for common data fields.

Figure 3.4b shows the MC88x00 architecture (Motorola 1988). It applied the Harvard architecture principle: A separation of datapath and instruction path to avoid internal pipeline stalls in the processor. The internal bus of the processor is separated in two source busses and one destination bus. The external data and instruction busses are each equipped with a Memory Management Unit, with data cache (virtual address side) and address translation cache.

The importance of these architectures for image processing is that, due to the various caches involved, a performance measurement cannot be based on counting assembler instruction but have to be measured using proper benchmarks.

A complete image scan usually does not fit in the data cache, operations within a neighbourhood do. However, image processing can be performed using a raster scan over the image, storing the pixels that are likely to change in a queue-like structure, followed by processing from an input queue structure to an output queue structure. If the queue structures are not too large they will fit in the cache, yielding a speed-up.

3.2.2 Vector processor architectures.

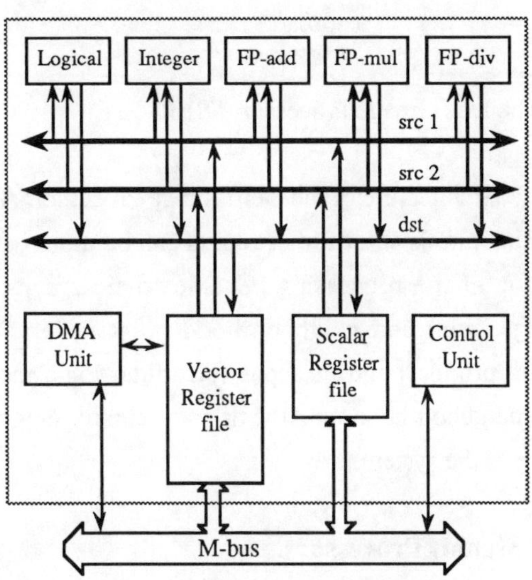

Figure 3.5 Generic set-up of a register vector processor.

In order to speed-up computations, specialized high speed vector processors can be coupled onto the M-bus. A general set-up of a vector register architecture is showed in figure 3.5. Usually, programs that operate on large data structures such as images, have a poor data cache performance. In such cases, vector processors may be fruitfully applied. Vector register machines copy entire vectors (e.g. a linear array of 64 to 256 integer or floating point numbers) from main memory to a register file, operate on it and store the results back under DMA in main memory. The older architectures, vector memory machines, directly operate on main memory. A drawback for image processing is that neighbourhood operations cannot easily benefit from the vector operations let alone more or less random walks through the image, for example in contour following.

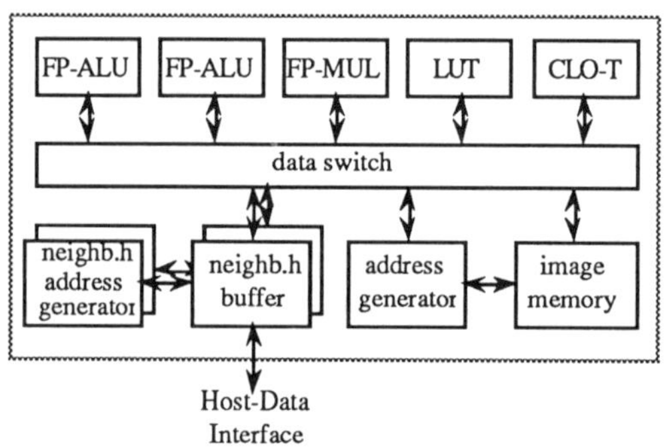

Figure 3.6 The Delft Image Processor (DIP).

Figure 3.6 shows the Architecture of the Delft Image Processor. Although the system is obsolete, its architecture is still of interest and can be more or less classified as a vector processor for image processing. Characteristic are its programmability, automatically image point and neighbourhood point generation, floating point capability (i.e. inner product) and its separate cellular logic module. The flexible addressing of large neighbourhoods and the use of recursive neighbourhoods are still a remarkable feature of the system.

3.2.3 Digital Signal Processors.

In applications involving dedicated image processing, one single Digital Signal Processors (DSP) often can be fruitfully applied to implement the solution. They can also be applied as a Processing Element for a programmable image processor. (see section 3.4.3). One of the modern DSPs with floating point capability is the

MC96002 (Motorola 1990), see figure 3.7. Characteristic is its dual bus interface which allows the DSP to be coupled onto external memory, AD-converters, a host system or to other neighbouring DSPs. Also characteristic is the separation of its internal bus into source and destination busses.

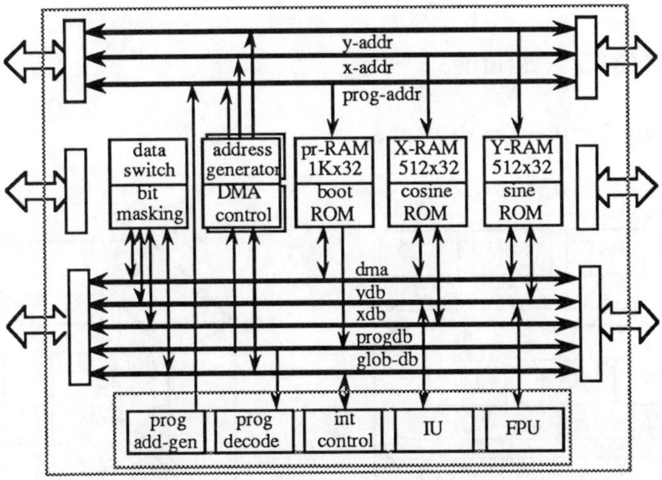

Figure 3.7 The MC96000, a single chip general purpose DSP with FPU.

Figure 3.8 shows a single board image processing DSP, the IPA-150 (Imaging-Technology 1990). It is part of a real-time pipeline system based on video busses.

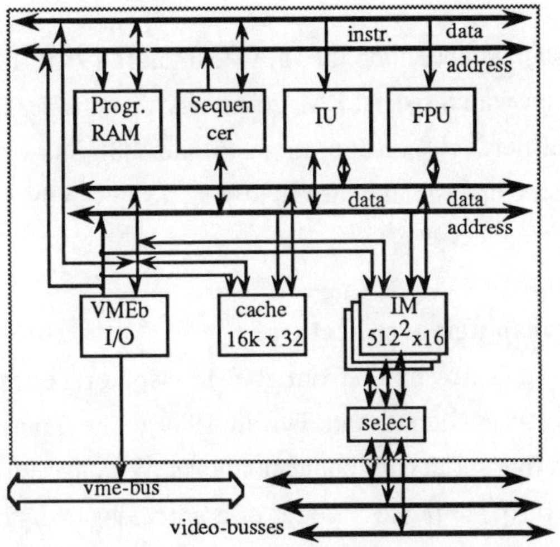

Figure 3.8 The IPA-150, a Single Board DSP for image processing.

Other modules of the pipeline system are video-acquisition board, ALU board, general grey-value filter board, morphological processing board and image memory board.

Characteristic of this DSP is its Harvard architecture, its on board image memory and its real-time pipeline connections.

3.3 MIMD systems.

3.3.1 The Ncube architecture.

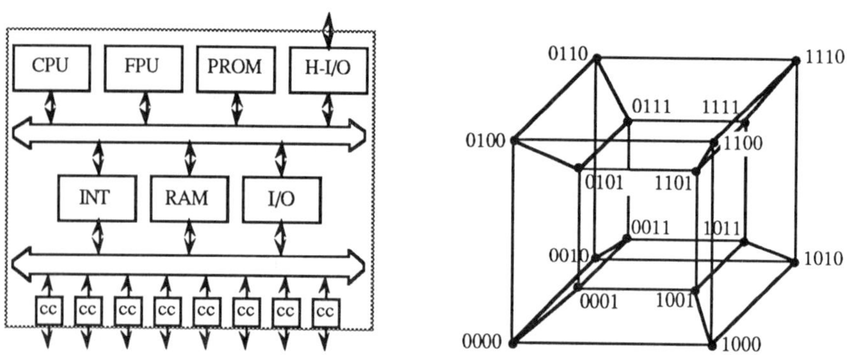

a) Ncube node

b) Ncube network topology

Figure 3.8 The iSPC, an Ncube node.

Figure 3.8a shows an Ncube, node the iSPC (Mokhoff 1985). It can be considered as a normal processor with 8 special DMA channels, that enables it to be coupled into a hyper-cubic data network configuration. Figure 3.8b shows a hypercube with dimension 4. Recent emphasis in hypercubic systems lies in the subject of pipelined routing hardware.

3.3.2 The Transputer architecture.

Figure 3.9 shows a T80x node from the Transputer series. (INMOS 1989, Homewood et al. 1987, Nicoud and Tyrrell 1989). The Transputer is generally characterized by its four serial communication links. With these links, configurations like a (4-connected) square processor grid or a processor tree structure, etc. can be made. An optional link switch chip makes it also possible to make connections under program control between more transputers. Remarkable is that after power-on a transputer can be booted from PROM or downloaded through one of its links by other transputers, so a host is not necessary. The Transputers are based on a parallel process paradigm, implemented with two process queues (high and low priority, 1 and 64 μsec.) and a hard-wired process scheduler. High priority processes run to completion, low priority processes can be pre-empted by a high priority process or its own time-slice end (2 msec) at jump or loop statements and another waiting process. A special event pin and two event timers with event queues makes the transputer competent for real-time control applications, e.g. in robotics.

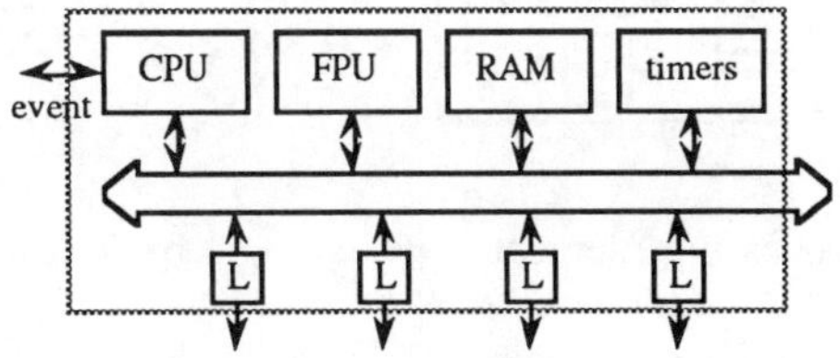

Figure 3.9 The T80x, a Transputer node.

Internal message passing between processes is implemented as a block copy. A writer puts a write pointer in a channel (memory word) and goes to sleep, a reader copies the buffer and awakes the writer. Note that the length of the message and the direction of the transfer is a mutual assumption between reader and writer. Message passing to other Transputers, using the external links, is performed identically using specific memory locations.

The Occam language (Inmos 1988) makes programming the transputer efficient, though also the C language is supported. Occam has special language constructs to handle processes, parallel execution of statements and priority event handling.

Through its synchronous 32 bits address and 32 bits data interface, external memory and devices such as AD converters, disk drives etc, can be connected.

For low-level image processing the transputers are not suitable due to their slow link speed (20 Mbit/sec raw speed). In the announced T9000 (Inmos 1991) this speed is increased to 100 Mbit/sec raw speed. Applicability of this device for image processing needs investigation.

3.4 SIMD systems.

3.4.1 The CLIP4 (SPA/LPA).

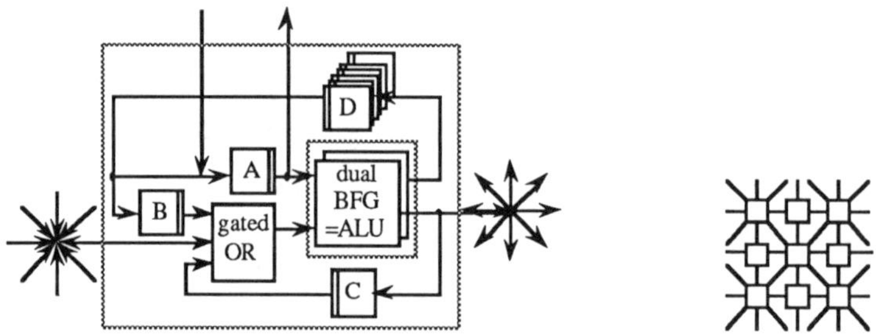

a) Processing Element b) Data network topology.

Figure 3.10 The CLIP-4 Square Processor Array.

The CLIP4 processing element (figure 3.10a) was used to build a Square Processor Array (figure 3.10b) (Duff 1982). It is a single bit Processing Element with connections to its 8 nearest neighbours. Greyvalue operations must be performed bitplane wise using the C(array) register. Both Monadic (A register) and Dyadic operations (A and B register) can be performed. The single bit ALU can be modified as two Boolean Function Generators (BFGs) in which one is used to store the resulting pixel and the other is used to transfer results to the neighbours. This, combined with a combinatorial path from neighbour input to neighbour output enables the programming of recursive neighbourhood operations. The neighbourhood can be accessed in parallel for cellular logic operations. Moreover, the recursive propagation over the image can be used as ripple-carry, thus allowing fast accumulative additions and subtractions of for example (row-size) vectors by (column-size) bits.

A bottleneck in the system is formed by the A register that is also used to shift data in and out the PE, hence disabling the possibility to pipeline array I/O and processing. A second bottleneck is the restricted number of internal storage (32 bitplanes due to the 32 bits D-register in the PEs). The CLIP4 can be considered as the archetype of a single bit SIMD processor for low-level image processing.

3.4.2 The Connection Machine (SPA/Ncube).

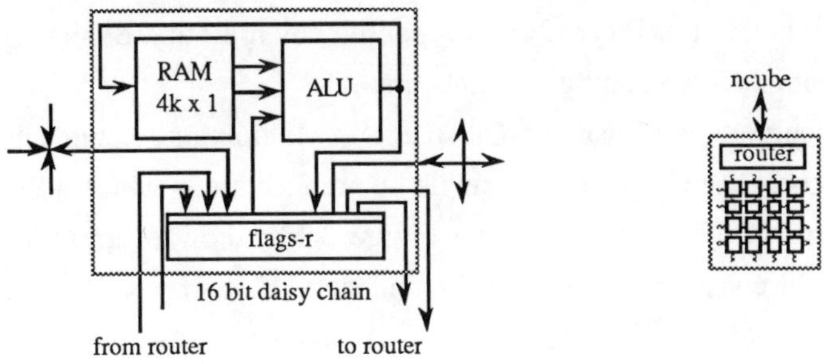

a) Processing Element b) Chip configuration

Figure 3.10 The Connection Machine.

Another known SIMD system is the Connection Machine. The PE is a single bit processor with internal RAM and ALU. The 16 PEs in one chip are 4 nearest neighbour connected but can be daisy-chained to perform 16 bit operations. Each chip has an on-board router to enable the chip to be a node in an Ncube configuration (with N ≤ 12). In this way data can be transferred to PEs on larger distances than nearest neighbour. The CM was designed with an emphasis on LISP processing.

3.4.3 The PICAP-3 and the Centipede (LPA).

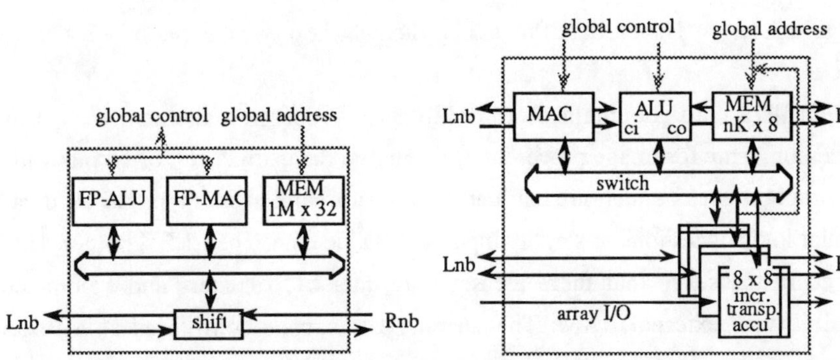

a) PICAP-3 Processing Element b) Centipede Processing Element group

Figure 3.11 Linear Array Processing Elements for Image Processing.

Two PE architectures specially designed for image processing are the PICAP3 (Lindskog 1988) and the Centipede (Schmidt and Wilson 1989). Both are designed to be configured in a Linear Processor Array configuration. Both use shift registers to communicate with their left and right neighbours.
The PICAP3 incorporates a floating point ALU and Multiplying Accumulator (inner product processor). The Centipede is a single bit architecture in which 8 PEs can be grouped to perform bytewise operations. Figure 3.11b shows a group of 8 PEs. Each group can be considered to have an 8 bit integer ALU and MAC. Characteristic for the Centipede is its 8 x 8 accumulators, that are able to transpose data and are equipped with an auto-increment possibility. Indirect addressing (address can be offset by the data) is possible in both systems. The PICAP3 was designed with an emphasis on 3D image processing, the Centipede on real-time 2D image processing.

3.4.4 The CLIP7a (LPA).

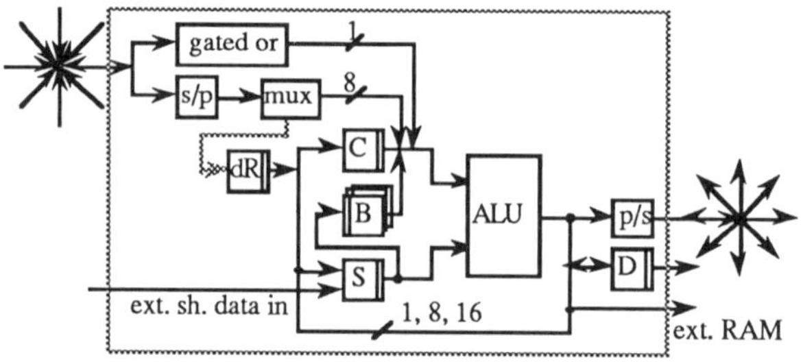

Figure 3.12 The CLIP7a Processing Element.

The CLIP7a chip (Fountain et al 1988b) is a Processing Element for a Linear Processor Array for image processing. Its internal datapath is 1, 8 or 16 bits wide. It has 8 neighbour connections that can be used to parallel input single bit data for cellular logic operations or serially input grey-value data. The CLIP4 bottlenecks are solved in the sense that there are separate data I/O registers and a connection possibility to external RAM. The 16 bit ALU can be configured as a Boolean Function Generator for cellular logic operations. The S(hift) register can be used to support the implementation of multiplications.
One of the most characteristic features of the PE is the "use CC" pin of the chip that can be used to switch from pure SIMD mode to a local autonomy mode. In the local autonomy mode the control of the PE can be modified locally with the C register.

This register may: Control the gated neighbourhood input OR, the neighbour selection and the B register address, specify a "use ALU" or "pass ALU" operation, set the Carry in value of the ALU and set a flag for the conditional load of the S and C register and the neighbourhood-out. The actual selected neighbourhood direction is stored in a separate register and can be used as data. Note that the ALU function, the RAM address and the D register load are always externally controlled.
The local autonomy (Fountain 1988a) can be used for example to let PEs individually operate on other bitplanes or images, or communicate with processors in other directions. The introduction of local autonomy is merely intended as a speed-up of known SPA operations than an approach to a new class of operations. Although research is performed that may result in new types of algorithms.

3.5 Pipelines.

3.5.1 The Warp.

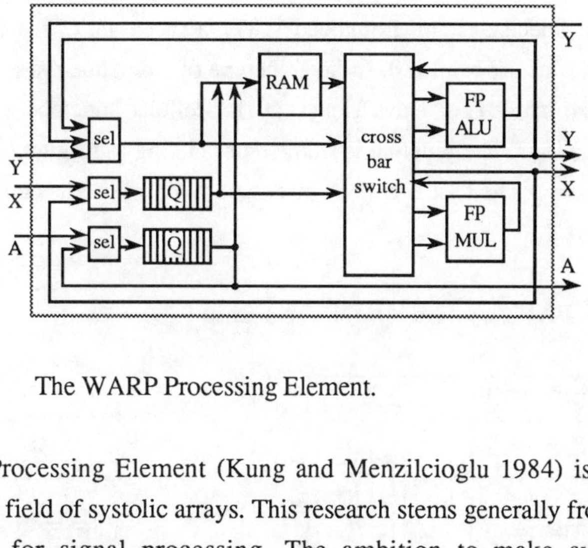

Figure 3.12 The WARP Processing Element.

The WARP Processing Element (Kung and Menzilcioglu 1984) is the result of research in the field of systolic arrays. This research stems generally from the field of VLSI design for signal processing. The ambition to make systolic arrays programmable for a field of applications rather than a single application has led among others to the design of the WARP. The WARP is a micro-programmable pipeline. Kernel of the PE are the floating point ALU and multiplier, connected to a crossbar switch to enable many datapath possibilities. The FIFO queues are used to compensate for the variable delays introduced by switching the datapath to different configurations. Pipeline registers enable synchronous connections to neighbouring PEs. An internal RAM can be used for the storage of coefficients for kernel operations. The software of the system facilitates programming of the array by aiding in algorithm decomposition to a systolic equivalent and generating microcode for a complete pipeline. Although not designed for image processing it is reported in literature that image processing can be performed on the system (Webb and Kanade 1986).

3.5.2 The Cyto HSS.

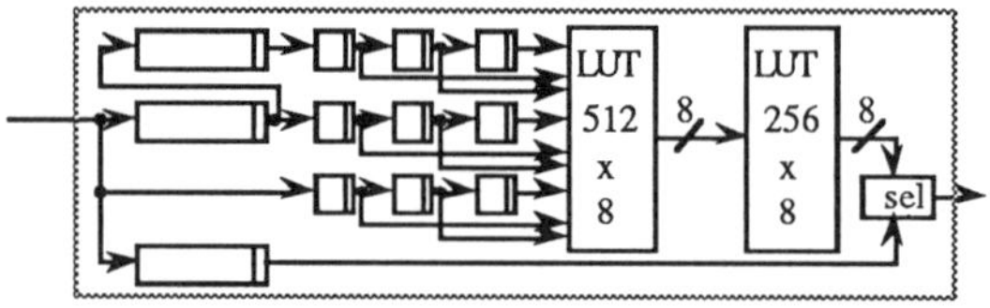

Figure 3.13 The Processing Element of the Cyto-High Speed System.

The Cyto-High Speed System (Lougheed 1987), a successor of the Cyto computer (Lougheed and McCubbrey 1980) is the archetype of a pipeline system composed of identical programmable PEs. The system is apt for cellular logic operations as well as greyvalue operations. The pipeline is frame recirculating (a ring topology including an image memory). The CLPE chip as discussed in chapter 5 has a similar, though extended design.

3.5.3 The PAPS system and Febris chip.

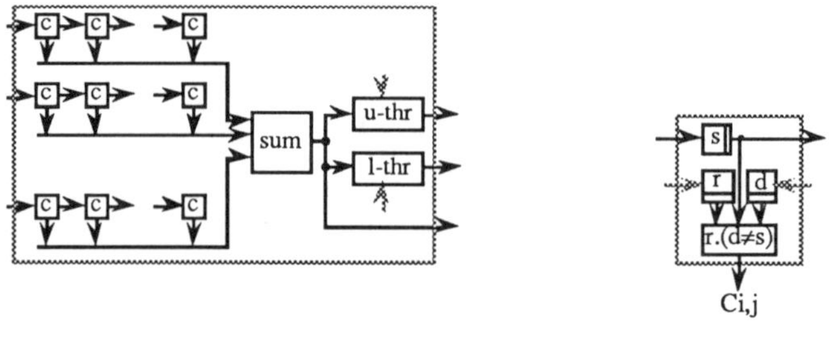

a) Febris chip. b) A pixel matching node.

Figure 3.14 A Febris Processing Element.

Many industrial pipeline systems are based on a video input card that generates a stream of pixels on one of a set of video busses. Each successive board on the bus contains a special processor (or a frame buffer) that can select one or more of the busses as input and one or more of the busses as output. In this way the pipeline set-up is made programmable, albeit in practice only at system configuration time. Usually the video input card yields the absolute timing of the pipeline system in the form of a horizontal and vertical sync signal. Each successive card in the pipeline yields a delay from a few pixels to a few image lines. In some industrial systems each card is able to compensate for the delay of the previous cards but in some systems this is not possible (sic!).

The Philips PAPS system (Thissen 1985) is a similar pipeline system, merely used in various in-house applications and within some commercial industrial apparatus. One of its cards, the Real-Time-Recognizer, a template matching device, was put into silicon. The result, the Febris chip (figure 3.14a) is able to match a 16 x 16 binary template with a binary image in video speed (Persoon 1986, 1988). The template matching can be used to detect local features in the image like edges, holes etc. Using Febris chips in parallel allows matching on several features in parallel e.g. identical features but in distinct rotations. Using Febris chips serially admits matching on complex structures, for example holes at specific positions on a circle, with a hole and a circle template in series. Within a template, foreground pixels, background pixels and don't cares may be specified (with registers r and d, figure 3.14b). The don't cares are used to compensate for small object rotations and noise at object edges. The results of the 16 x 16 pixel matches are summed and this result is compared with two thresholds. One threshold can be used to specify the number of foreground noise points that are tolerated and the other threshold can be used to specify the number of background noise points that are tolerated.

3.5.4 The Sarnoff Pyramid chip.

Multi resolution image processing can be performed with pyramidal architectures. An example of a full size pyramid, consisting of a stack of decreasing sized 4-connected Square Processor Arrays is the PAPIA (Cantoni and Levialdi 1987). Each processing element is physically connected to four sons in the plane below and to one parent in the superior plane. Each plane can run an independent SIMD program.

An example of a real-time pipelined pyramid is the Sarnoff Pyramid Vision System (Burt and van der Wal 1987, 1990) and its recent partially implementation in the

Sarnoff Pyramid Chip (van der Wal 1991) (see figure 3.15a). The kernel of the Sarnoff Pyramid chip forms a 5 x 5 binomial filter section with 1K shiftregisters to create the neighbourhood. The binomial filter can be programmed to compute Gaussian, Laplacian, gradient, moment and sub-band pyramid transforms and its inverse transforms.

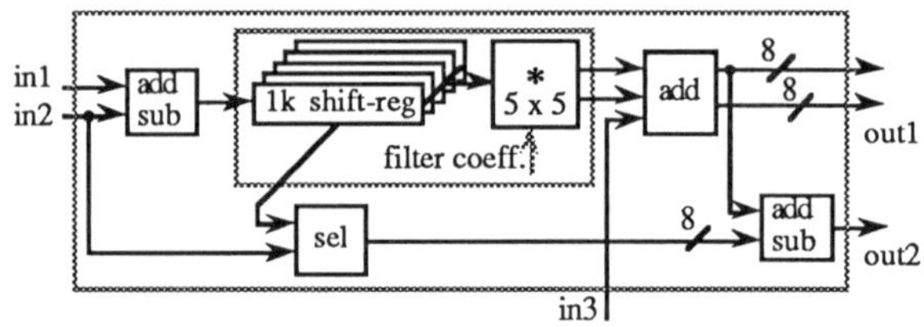

Figure 3.15a The Sarnoff Pyramid Chip.

The chip has automatic image edge control by either extending the data at all image edges or substituting a constant value. The chip operates on an 8 bit input and produces a 16 bit result. For increasing precision two pyramid chips can be connected in parallel, using the *in3* connection (see figure 3.15a). The chip automatically delays the horizontal and vertical video sync signals for smooth assembling of video pipelines. The chips can be used either in a cascade, each chip operating at half the clock rate of the previous chip, or be used with a down or up sampling framestore (see figure 3.15b and c). Figure 3.15b shows the simultaneous computation of the Gaussian and Laplacian (Difference of Gaussian) pyramid using a downsampling framestore. Figure 3.15c shows the reconstruction of the original image from the Laplacian pyramid using an upsampling framestore.

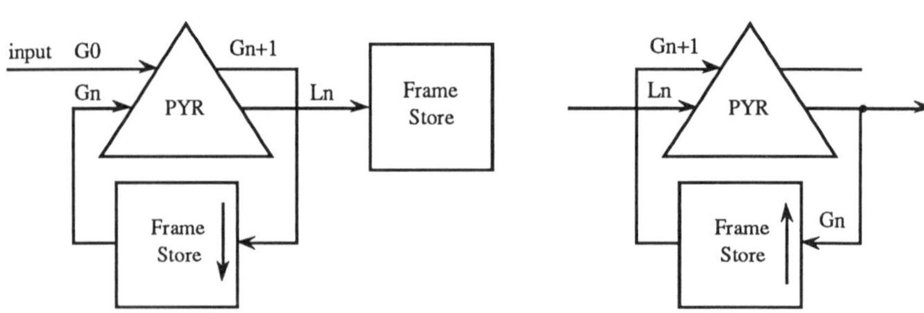

b) Gaussian and Laplacian pyramid transform c) Inverse transform

Figure 3.15 Implementation of Pyramid transforms.

3.5.5 The NEC IMPP.

A quite different design albeit still a pipeline design is the NEC-IMage Pipeline Processor (IMPP) (NEC 1985, Iwashita et al. 1986). This Processing Element is based on the dataflow paradigm. Figure 3.16a shows a dataflow graph of the function $f(x) = x^2 - 2x + 3$. The flowgraph can be envisioned as an asynchronous pipeline running from left to right with parallel branches. Data can be input at the terminal on the left and leave the system at the terminal on the right. Each block in the flow diagram performs an operation. Note that stepwise refinement is possible in the paradigm. Function blocks can be refined until a basic block level, the machine instructions. Each block is assumed to have FIFO queues at its input branches that may store data tokens and hence uncoupling between the blocks is achieved. Each function block can be envisioned to be a separate processor, but it is also possible that one or two processors connected to a mutual data network implement the flowgraph.

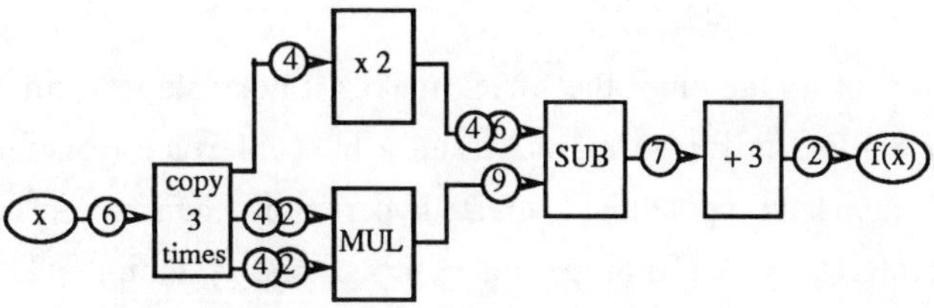

Figure 3.16a Dataflow graph of the function $f(x) = x^2 - 2x + 3$.

This can be achieved by storing at each token in addition to the data the identification of the branch it is waiting on. If the token enters an arbitrary processor, the processor knows that he is the one that implements the function belonging to the branch, processes the token, generates an output token with a new branch identification and puts the token onto the network. For function blocks with more inputs, the processor must, if necessary, temporarily park the token and wait for a buddy token that is awaited on the other input of the function block.

Figure 3.16b shows the IMPP token ring processor that uses this principle. Tokens that enter the system are checked on their flow-graph branch identification (ID). If this PE is not programmed to execute the function belonging to that branch, the token is passed to the output. If the token is programmed to be processed in this PE, the token is passed through the link table. In the link table, the branch-ID is swapped for the output branch-ID of the flowgraph block. In the function table the operation code for the processing unit is connected to the token. The data memory is used for waiting partner tokens or retrieval of constants. Via a process queue the token is processed in the processing unit. Some operations require a second pass through the internal ring, if not, the resulting token is passed to the output.

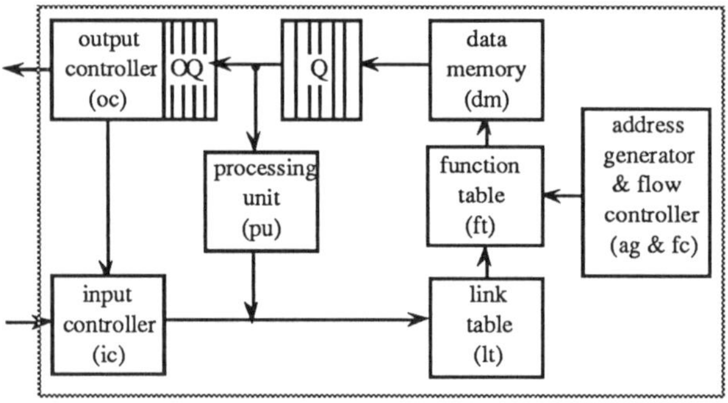

Figure 3.16b The NEC Image Pipeline Processing Element (IMPP).

The availability of a glue chip, the MAC, makes it possible to form a ring from a maximum of 8 PEs, an image memory and a host interface (together a cluster). Recent MAC chips have a potential to create hyper-cubic ring structures as each MAC has two ring interfaces. A prototype image processing system, the TIP4 was built, consisting of one ring with 8 clusters, i.e. 64 IMPPs (Fujita et al. 1990). The IMPPs do not have floating point properties.

3.6 Evaluation and discussion.

The virtual memory in the uni- or multiprocessor systems gives each user virtually an infinite memory. The locality of reference principle (temporal, spatial and sequential) makes it possible to hide the virtual character of the memory to the user. Temporal locality by the data cache, spatially by the address translation cache and sequentially (on a macro scale) by the next page prediction of the paged memory management system. The sequential locality of reference on a micro scale can, optionally, be exploited by vector processors. The locality of reference principle can be applied profitably on any information stream, as can be seen in the MC88x00 system with its radically uncoupled streams, such as sources, destination, instructions and data.

The parallel with image processing systems is that, if possible, the user would like to operate on many images of infinite size, using many processors in parallel.

It is obvious that for low-level image processing, the neighbourhood addressing has a spatial locality of reference, the kernel coefficient retrieval a temporal, and the traversal through the image points a sequential locality of reference.
Practically all systems for low-level image processing are based on fast and adequate handling of the kernel and the attendance of all points in the image. Systems not specially designed for image processing (DSP 96000, The Connection Machine, The Warp) go wrong in the details, like adequate image edge handling, parallel retrieval of the neighbourhood, special treatment of cellular logic operations. The latter is solved very nicely in for example the CLIP7a and the Centipede. With cellular logic operations the entire neighbourhood can be principally processed in parallel, arithmetic processing in a neighbourhood is essentially sequential. The solution sought to overcome this is in the data parallelism of the nearest neighbour connected Processor Arrays. Most greyvalue filters can be decomposed in iterations of 3 x 3 filters, or even 3 x 1 and 1 x 3 filters. For the handling of larger irregular neighbourhoods, however, nearest neighbour connected machines become bothersome. Large sparsely filled kernels or template matching are examples of this. Systems like the DIP, with a large neighbourhood cache, or the IPA do not struggle with this. If indirect addressing is possible, histograms can be formed using a part of the image data structure.

The difference in application field often leads to a choice for SIMD systems or for Pipelines. The nice property of an SIMD system is that it programs more or less as flexible as a sequential computer. The nice property of a pipeline is that, if made real-time, every processing step can be monitored in real time. Especially in industrial environments, one should not underestimate the human capability to detect potential problems (e.g. noise) and to solve these problems by looking at the live images. Moreover, pipelines are very suitable for Recursive Neighbourhood operations.
For proper real-time operation "details", such as the PE delay compensation problem, the image edge handling, etc. should likewise be solved correctly, as in the PYR and the Cyto-Hss. The Centipede is an example of an SIMD system in which solutions were explicitly incorporated for real-time operation. A separate I/O shift register can take in a stream of pixels from a video acquisition unit and provisions have been

made to form a pipeline of LPAs if a single LPA is not fast enough for real time processing .

However, as soon as the filtering domain is left, the benefit of the current image processing systems becomes less. Then, the structure and the granularity of the data is different and hence the data structures and the processor assignment should be different. Structures in the image are for example objects, object edges, lines, curves or areas in images. The locality of reference principle now encourages that the involved data structures are cached and processors are assigned to each separate embodiment of the new data structure in the image. Beware that the lowest data entities in the new data structure are still image points to be retrieved from the image data structure! (A parallel with data cache and main memory looms up).

To handle coarse grain superstructures in the original fine grain image data, for processor arrays, solutions are sought in establishing connections over larger distances between the PEs, or in local autonomy. However, this leaves the problem untouched of the many PEs that are idling if they are not involved in the new structures of the data.

Note that at this point pipelines are extinct because they are both not very flexible in programming and can hardly make any connections over larger distances between PE's. With as a notable exception the token ring of the IMPP, in which tokens shift automatically to a PE that is addressed by the address accompanying the data in the token.

MIMD systems do not give any relief, because they have no special provisions for low-level processing and their communication bandwidth is usually to low. Although the MIMD principle starts to become applicable as more data dependent control is involved when going to intermediate level image processing. Hybrid systems are proposed and applied (e.g. Burt and van der Wal 1987; Pyramid pipeline and MIMD system) to overcome this. The interface between the low-level machine and the intermediate level machine may, however, form a new bottleneck.

A nice property of the Transputer line of MIMD systems is, that it is based on a clear paradigm, co-operating parallel processes, and that the architecture is strongly supported by system software that eases programming in the new paradigm (excluding the infamous folding editor). The fact that a parallel solution will run on a single Transputer and will run, without adaptation, faster as transputers are added, has the same appeal as the virtual memory concept.

The IMPP is also based on a clear paradigm, dataflow processing, which is reasonably supported by software. The nice property is that the parallel and pipelineable steps follow directly from the flow diagram of the solution and are as such recognized and executed by the system. As with the transputer, the implementation will run on a single processor and performance improves when adding more processors, without adapting the implementation. However, dataflow processing becomes cumbersome if the data does not need to flow (Dekker et al 1987, Jonker et al 1990a), for example at algorithms based on histogram search (Huang 1979). Pure dataflow processing leaves not much room for parallel operations (cooperating, mutually excluded processes) on large data structures. Moreover, a bottleneck in the IMPP is formed by the fact that PE interconnections and memory addressing takes place over the same ring.

Concluding:

The memory management systems of current uni/multiprocessor systems are not optimally tuned for image processing. This cannot be solved by using vector processors.

Special purpose systems to boost the performance of a uni/multiprocessor system must be explicitly designed for image processing, properly solving the few peculiarities of this field. The same applies if real-time performance is demanded.

Low-level image processing systems are too much bound at the image-data and filter-kernel structure to allow flexible introduction of other data structures, necessary for intermediate level image processing.

Coupled MIMD and SIMD systems must be carefully designed, not leaving bottlenecks as the finest detail of the datastructures for intermediate level algorithms are still pixels.

Virtual systems (virtually unlimited bounds), with a single clear paradigm, well supported by system software are attractive. The attractiveness stems largely from the fact that the system leaves the programmer free within a paradigm and tries to predict the behaviour of the program, resulting in the optimizing of its own configuration (caching). Rather than letting the programmer optimize his program into a fixed datastructure. (Then we might as well program in assembler on a sequential machine.)

And, virtual systems certainly are necessary when the number of image elements are too large for the number of processors available.

4 Morphology based image processing.

The goal of this book is to discuss research in the field of the design of fast computers for image processing. After the introductions of chapter 2 and 3, the next chapters are devoted to the design of computer architectures for image processing, with the mathematical morphology as a starting point. Chapters 4 and 5 are directed to low-level image processing. Chapter 6 is devoted to extensions towards three-dimensional and intermediate level image processing.

In the field of low-level image processing, many special hardware systems are based on binary image processing using single bit Processing Elements. Provided with a carry-bit, these systems can be used for greyvalue processing by repeatedly processing single bitplanes, with the result that less powerful PEs can be used and massively parallel processing becomes possible.
The field of low-level image processing based on single bit operations in a small neighbourhood is called Cellular Logic Processing. If architectures are founded on single bit processing, the field of Cellular Logic Processing need to be first researched intensively. The result of such a study is presented in this chapter.

In section 4.1, starting from a definition from the mathematical morphology, Cellular Logic Processing is discussed on a basis of inexact matches of a set of 3 x 3 binary structuring elements with a local neighbourhood in the image. This section has an informal, practical approach and can be seen as an introduction to the next section (4.2) which gives a more detailed presentation of research on mathematical morphology in multi-dimensional images. The architecture based on section 4.1 is described in chapter 5.

The aim of section 4.2 is the issue of automatic generation of structuring elements for topology preserving thinning, or skeletonization, of images with a higher dimension than two. The outcome of this is an onset to a theoretical framework for high dimensional mathematical morphology, plus a thinning scheme and method that can be used to constitute an architecture for three dimensional image processing. The architecture based on this section is described in chapter 6.

In chapter 5, using the knowledge of section 4.1, a hardware architecture for real-time low-level image processing is described, including its implementation in VLSI technique. The architecture is based on a pipeline of identical Cellular Logic Processing Elements, that are able to execute the cellular logic operations as developed in section 4.1. By stacking *b* Processing Elements on top of each other, *b*-bit greyvalue processors can be obtained. These processors can be cascaded to form a pipeline for real-time low-level greyvalue-image processing for two dimensional images.

Finally, in chapter 6, a theoretical design of an image processing system is described, that supports both three-dimensional and intermediate-level image processing in addition to two-dimensional, low-level image processing. Cellular logic processing is also the basis for this system, but now for two *and* three-dimensional images, using the cellular logic technique that is elaborated in section 4.2. Both the concepts of highly parallel and real-time processing are abandoned in this architecture. It is a hybrid architecture, in the sense that it can be programmed to behave as a Linear Processor Array, a Pipeline or a Token Ring processor. The special processing facilities for intermediate level image processing in a PE are bucket sorting queues. A firm interface with an MIMD system is included in the design.

4.1 Cellular logic processing.

4.1.1 Introduction.

In this section an informal and practical introduction to Cellular Logic Processing is presented. The method described here has been developed in parallel with an architecture for real-time processing of binary images, implemented in the CLPE chip (see chapter 5). The method supports the architecture but is by itself entirely applicable on other machines.

In the first sub-section, starting from the mathematical morphology, a description of Cellular Logic Operations as Hit-or-Miss transformations with sets of masks is presented.

In the next subsection, the commonly used Cellular Logic Operations are discussed. A separation is made in Local Neighbourhood -Cellular Logic- Operations (LNOs) and Recursive Neighbourhood -Cellular Logic- Operations (RNOs). The topic of topology preserving thinning, or skeletonization is highlighted.

The next sub-section presents some skeleton variants, that are inherent to the mask-set concept, as developed in the preceding sub-sections. Section 4.1 ends with conclusions.

4.1.2 Mathematical Morphology.

The field of mathematical morphology is described in several works (Golay 1969; Preston 1970; Serra 1982; a clear explanation can be found in: Giardina 1988).

One of the basic operations in mathematical morphology is the Hit-or-Miss transformation. Most architectures in this thesis are founded on this transformation.

As a starting point we will quote the definition by (Serra 1982):

The Hit-or-Miss transformation is a point-by-point transformation of a set X that is performed in the following way:

Choose and fix a structuring element S; the datum of S being two sets S^1 *and* S^2*. Suppose S is centered at the point x of X, denoted by* $S = (S_x^1,\ S_x^2)$ *then a point belongs to the Hit-or-Miss transformation* $Y \leftarrow X \otimes S$ *of X, if and only if* S_x^1 *is included in X and* S_x^2 *is included in the complement* X^c *of X, or:*

$$Y \leftarrow X \otimes S \equiv \{x \mid (S_x^1 \subset X,\ S_x^2 \subset X^c)\} \tag{4.1.1}$$

This general definition in the mathematical domain will be applied in the physical domain of image processing. We will bring definition (4.1.1) in correspondence with the definitions in section 2.2.1 in the following way:
Let us associate with the universe -the union of X and X^c-, a square tessellated image X of size m x n, and with the elements of the universe, the pixels of image X. Then we can associate with the set X, the foreground of image X and with the set X^c its background. Likewise for the sets Y and Y^c, and image Y.
As we will apply the transformation in this thesis on binary images only, we will define the image foreground to have the value TRUE or 1 and the image background to have the value FALSE or 0. Likewise for Y.
With the transformation of input set X to output set Y using structuring element S, a transformation of input image X to output image Y using structuring element S can now be associated. Like X and Y, S can for instance be implemented as an image.
As structuring element S consists of two disjunct sets S^1 and S^2, then the universe is formed by the union of the disjunct sub-sets S^1, S^2 and S^c, S^c being the complement of the union of S^1 and S^2. Let us associate with the union of S^1, S^2 and S^c, a square tessellated image S of size i x j, and with the elements of S^1, S^2 and S^c , the pixels in the image S. Then we can associate with the set S^1 the foreground pixels, with the set S^2 the background pixels, with the union of S^1 and S^2 the do-care pixels and with the set S^c the don't care pixels. In our case of binary image processing, the foreground is defined to have the value TRUE or 1, the background to have the value FALSE or 0, and the don't cares to have the value D.
The don't cares are introduced to implement the unrestricted set S with an image S with a fixed size and shape.

Let x, y and s be the elements (pixels) of X, Y and S, and let the size of S e.g. be i = j = 3, then the 3 x 3 structuring element S consists of s^0, s^1,...s^8; the east (E), north-east (NE), north (N), ... , south-east (SE) neighbours of the central (C) pixel s^8 itself, and an equivalent 3 x 3 neighbourhood M_k around pixel x_k, consists of x_k^0, x_k^1,...x_k^7: the E, NE, N, ... ,SE neighbours of the pixel x_k and x_k^8 the central (C) pixel x_k itself.

The point-by-point transformation $Y \leftarrow X \otimes S$ can now be implemented for binary image processing with the neighbourhood transformation $Y \leftarrow X \cong S$, or

$\{\forall k: y_k \leftarrow M_k \cong S\}$ with M_k and S centered around pixel x_k, and with $\cong$ the symbol for an inexact match.

Informally, the inexact neighbourhood matches $\{\forall k: y_k \leftarrow M_k \cong S\}$ can be described as follows: If for any pixel x_k in an input image X its 3 x 3 neighbourhood M_k matches inexactly with a 3 x 3 structuring element S, the pixel y_k of output image Y is set to one. If M_k does not match with S, y_k is set to zero. Note, that in the inexact neighbourhood match, the position of the foreground pixels in S should match with the position of the foreground pixels in M_k, the background pixels in S should match with position of the background pixels in M_k and in the don't care positions of S a match is not required. This is depicted in figure 4.1.1

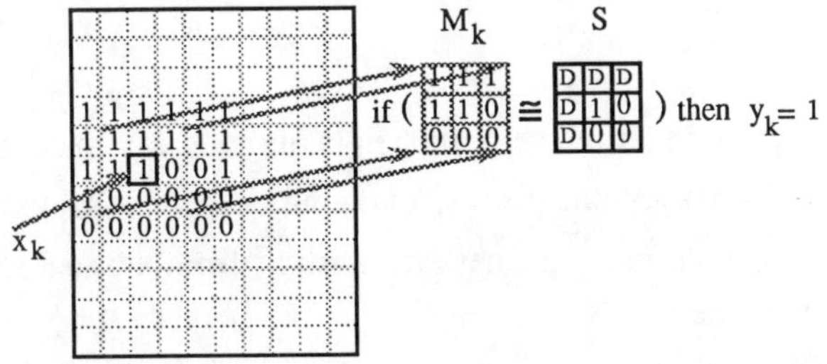

Figure 4.1.1 The inexact neighbourhood match.

More formally, this inexact neighbourhood match $y_k \leftarrow M_k \cong S$ is described by:

$$y_k \leftarrow \bigwedge_{i=0}^{8} \left(x_k^i \cong s^i\right) \tag{4.1.2}$$

Where $(x \cong s)$ is defined as: $(x \tilde{\wedge} s) \tilde{\vee} (\bar{x} \tilde{\wedge} \bar{s})$

with: $\bar{x}$ the complement of x, etc.

(4.1.3a)

and $(x \tilde{\wedge} s)$ and $(x \tilde{\vee} s)$ be defined as:

x	s	$x \tilde{\wedge} s$
0	0	0
0	1	0
0	D	0
1	0	0
1	1	1
1	D	1

x	s	$x \tilde{\vee} s$
0	0	0
0	1	1
0	D	0
1	0	1
1	1	1
1	D	1

(4.1.3b)

The definition of the inexact neighbourhood match may be extended in two ways (Jonker and Duin 1985), first to:

$$y_k \leftarrow \bigvee_{i=1}^{p} \left(M_k \cong S_i^s\right), \tag{4.1.4}$$

or: If for any pixel x_k in an image X its neighbourhood M_k matches one mask S_i^s from a given set of masks S^s, the pixel y_k of output image Y is set to one (else set to zero). Note, that this means that the logic OR of all individual matches is taken.

And further to:

$$y_k \leftarrow \left(\bigvee_{i=0}^{p} \left(M_k \cong S_i^s\right)\right) \vee \left(\bigvee_{i=0}^{q} \overline{\left(M_k \cong S_i^r\right)}\right), \tag{4.1.5}$$

or: If for any pixel x_k in an image X its neighbourhood M_k matches one mask S_i^r from a given set of masks S^r, the pixel y_k of output image Y is set to zero (else set to one), but if its neighbourhood matches one mask S_i^s from a given set of masks S^s, the pixel y_k is set to one.
The image transformation $Y \leftarrow X \cong S$ can now be implemented with a structuring element, implemented as a set of masks S, that consists of a subset S^s (the SET-masks) and a subset S^r (the RESET-masks), one of which may be empty.
This means that either a pixel y_k is set to zero, if one of the RESET masks fits, or the pixel is set to one, if one of the SET mask fits, where the SET masks dominate over the RESET masks.

Finally, a 'mask-image' Z can be used to locally disable the transformation[1] , using:

$$y_k \leftarrow \overline{\left(z_k \cong z\right)} \wedge \left(\left(\bigvee_{i=0}^{p} \left(M_k \cong S_i^s\right)\right) \vee \left(\bigvee_{i=0}^{q} \overline{\left(M_k \cong S_i^r\right)}\right)\right) \tag{4.1.6}$$

---

[1] See also chapter 2. To call Z a 'mask-image', is rather confusing at this point. The name stems from the fact that with Z it is possible to locally mask-off the operation (data dependent). It has no connection with the hit-or-miss masks that implement the structuring element. However, traditionally Z is called the mask-image; and the bit to enable/disable the mask image operation, the 'mask-bit'. To change that at this moment would give more confusion.

If a mask-bit z is set don't care, the transformation is always enabled. If z is do-care, then if the pixel z_k of Z matches with the mask-bit z, the operation is disabled, else enabled. In the sequel, two operations are shown that use Z: The propagation operation and the anchor-skeleton.

Note that the word structuring element is used in both the mathematical case ($\mathcal{S}$) as in the binary image processing case (S). For the binary image processing case we will also use the word mask (indicated with S as well). In the sequel we will extend the neighbourhood match ($\cong$), from a match with one single mask to a match with a set of masks (indicated with S_i). We will also show that in many cases a mask set $\{S_i \mid (i > 1)\}$ can be transformed into a set with a single mask (S_1), and vice versa, without affecting its functional properties. As the distinction between the function of a single mask and a set of masks often cannot be made, we will refer to both as being a structuring element, if we want to emphasize the operation that the mask or the set of masks performs. For example, exactly the same erosion can be performed by a single mask and by a set of masks. In both cases we will refer to those as 'the structuring element for the erosion' if we want to address the functionality.

Using this concept of the Hit-or-Miss transformation, a class of operations on binary images, the cellular logic operations can be described. A cellular logic operation on an image X can be done by performing the Hit-or-Miss transformation once:

$$Y \leftarrow X \cong S, \tag{4.1.7a}$$

twice or more (the n-pass form):

$$Y \leftarrow ((X \cong S) \cong S)...), \tag{4.1.7b}$$

or infinite (the iterative form):

$$X_0 \leftarrow X \,;\, \{\forall i\colon X_{i+1} \leftarrow (X_i \cong S)\} \,;\, Y \leftarrow X_\infty \tag{4.1.7c}$$

In the latter case an iterative scheme is used, which can be implemented practically by transforming the image until it is idempotent under the transformation.

In many cases practical use can be made of the intermediate results in the output image. This is called (spatial) recursion (see also chapter 2).

For instance, if the transformation is performed by a raster scan over the image, i.e. from top-left to bottom-right, the fact that the neighbours NE, N, NW, W of the pixel y_k have already obtained a new value can be utilized. For this purpose the neighbourhood M_k of pixel x_k will be extended by:

$$y_k^{r1}, y_k^{r2}, y_k^{r3}, y_k^{r4},$$

the newly obtained values, and the structuring element S will be extended by:

$$s^{r1}, s^{r2}, s^{r3}, s^{r4}.$$

So now both the Local Neighbourhood (LN: the original value x_k and its neighbourhood $x_k^0, x_k^1, \ldots x_k^7$), and the Recursive Neighbourhood (RN: the newly obtained values $y_k^{r1}, y_k^{r2}, y_k^{r3}, y_k^{r4}$), can be used in the neighbourhood matching procedure.

The class of cellular logic operations performed with this extended neighbourhood will be referred to as *recursive neighbourhood operations* (RNO) in contrast with the *local neighbourhood* operations (LNO) that use the normal neighbourhood only.

4.1.3 Cellular Logic Operations.

Based on the concepts of the previous sub-section, various Cellular Logic Operations can be distinguished. The sets of masks that are used to perform the Cellular Logic transformation (4.1.6) from input image X (and Z) to Y can be appropriately drawn. Figure 4.1.2 shows the drawing conventions for mask sets that will be used throughout this thesis.

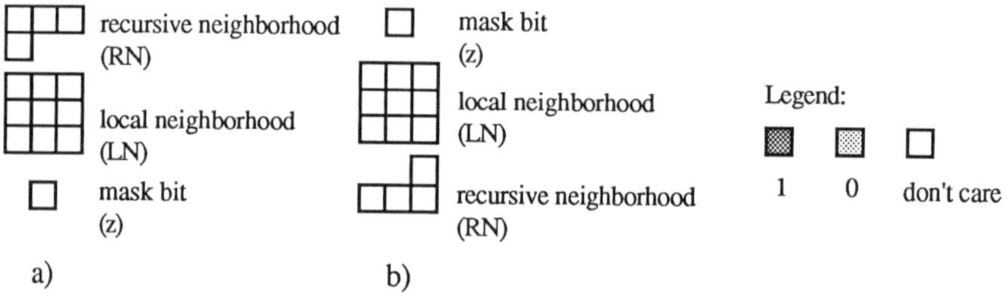

Figure 4.1.2 Mask set drawing conventions

It shows the outline of a single mask from a set. The recursive neighbourhood (RN) represents the pixels $s^{r1}, s^{r2}, s^{r3}, s^{r4}$ in the mask that should be matched with the pixels x_k^{r1}, x_k^{r2}, x_k^{r3}, x_k^{r4} in the image. The local neighbourhood (LN) represents the pixels s^0, s^{r1}, ...s^8 in the mask that should be matched with the pixels x_k^0, x_k^1,...x_k^8 in the image. The mask-bit (z) denotes whether the central pixel z_k of an optional mask image (4.1.6) is allowed to disable the operation (if it is do-care). The legend shows how foreground (dark grey: 1), background (light grey: 0) and don't cares (blank) are indicated **throughout this thesis**. Note, that when raster scanning the image from top-left to bottom-right, the pixels on the neighbourhood positions NE, N, NW and W, form the recursive neighbourhood; figure 4.1.2a, and when raster scanning the image from bottom-right to top-left the pixels on the neighbourhood positions E, SE, S and SW form the recursive neighbourhood; figure 4.1.2b).

The most frequently used known *normal neighbourhood* cellular logic operations are: Erosion, Dilation, Contour extraction, Spot noise removal (pepper and salt removal) and Majority vote (binary rank filtering). These operations wil now be described with mask-sets paradigm of sub-section 4.1.2.

Note, that within a square tessellated two-dimensional image with objects on a background, there are two possibilities for the connectivity of object and background pixels. The objects can be chosen to consist of pixels connected with one or more of their 8 neighbours at (E, NE, N, NW, W, SW, S, SE) or they can be chosen to be 4-connected, with one or more of their neighbours at (E, N, W, S).
See also (Rosenfeldt and Pfaltz 1966) for the connectivity paradox.
Consequently, in the first case the background is 4-connected, in the latter case it is 8-connected. The use of square tessellated binary images and the use of structuring elements of size 3 x 3 limits the possible metric of the operations; either a 4-connected -or city-block- metric or an 8-connected -or chessboard- metric (See also section 4.2).

Some mask-sets for simple cellular logic operations are (see figure 4.1.3):

- Copy the image (no-operation).
- Move the objects in the image one position to South West.
- Erode an 8-connected contour from the objects in the image.
- Erode a 4-connected contour from the objects in the image.
- Dilate a 4-connected contour to the objects of the image. This mask shows clearly the duality between a dilation of a 4-connected contour to foreground objects and the erosion of an 8-connected contour from the background around the objects.

- The detection of an 8-connected contour; The first mask of the set deletes all non 8-connected contour pixels of objects, the second mask copies the background. This mask set shows clearly a duality between contour detection and contour erosion.

Note that on top of each mask in figure 4.1.3. is indicated to which subset the mask belongs: Subset S^s (the SET-masks) or subset S^r (the RESET-masks). Below the mask the name of the operation is given. Masks are grouped to a set in the figures by a bracket. For example: The erosion 8 operation (erodes 8 connected contours from objects) is performed by scanning the input image with the third mask from figure 4.1.3. If the mask locally matches a 3 x 3 neighbourhood in the image the output pixel in that point is **set** (to foreground: 1). The last two masks of figure 4.1.3. form a set that extracts the 8 connected contours of objects in an input image. If one of the masks of the set locally matches a 3 x 3 neighbourhood the output pixel in that point is **reset** (i.e. set to background: 0).

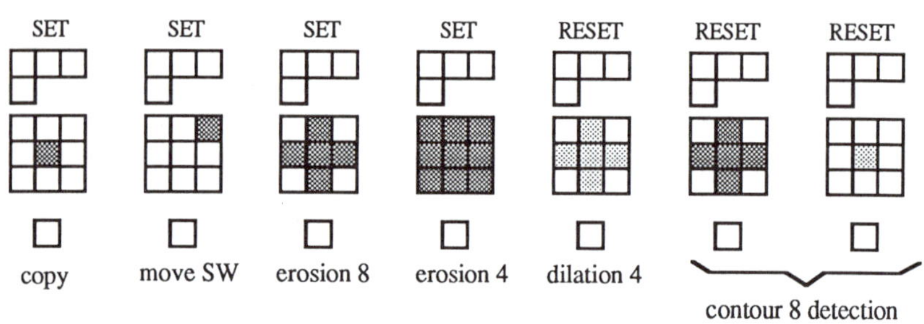

Figure 4.1.3 Some normal cellular logic operations.

The relation between the SET and RESET forms of the mask-sets is given by the switching algebra or Boolean algebra, i.e. the theorem of De Morgan (1847)[2]. Figure 4.1.4 shows that defining mask sets using only SET conditions is possible, although in some cases this is not very parsimonious. Equation 4.1.8 shows the transformation of dilation_4 from the disjunctive canonical form to the conjunctive canonical form.

2 see e.g. (McClusky 1965) or (Blaauw 1976 app. D: switching algebra).

$$\left(\left(\overline{C}\wedge N\right)\vee\left(\overline{C}\wedge W\right)\vee\left(\overline{C}\wedge S\right)\vee\left(\overline{C}\wedge E\right)\ \vee C\right)\equiv\left(N\vee W\vee S\vee E\vee C\right)\equiv\left(\overline{\overline{C}\wedge\overline{N}\wedge\overline{W}\wedge\overline{S}\wedge\overline{E}}\right)$$

(4.1.8)

This result is visualized in the figures 4.1.4 (disjunctive form) and 4.1.3 (conjunctive form). Dilation_4, the fifth mask of figure 4.1.2 expands to a set of five masks (figure 4.1.4) and Contour_8 , the set consisting of the last two masks of figure 4.1.3, transforms to a larger set (4 masks) in figure 4.1.4.

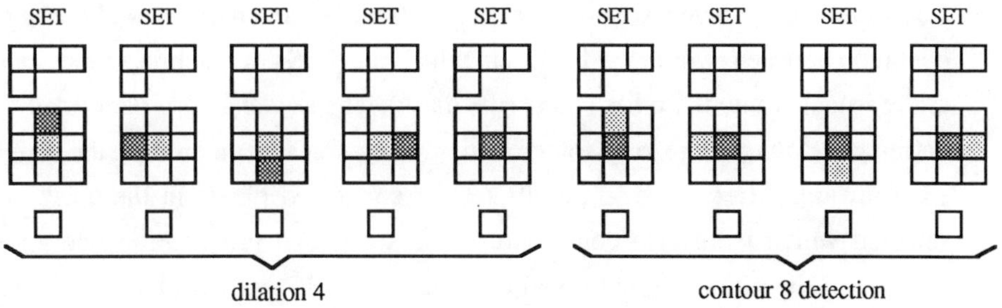

Figure 4.1.4 Some non-parsimonious mask sets for simple operations.

This leads us to the observation, that mask sets can be minimized by transforming them from the disjunctive canonical form to the conjunctive canonical form, or vice versa.

Proceeding with some more complex Cellular Logic Operations:

- Majority vote or binary rank filtering is a group of local neighbourhood operations that sets a pixel to the value of the majority (m) of its neighbourhood. The number of masks to implement a 3 x 3 binary rank filter is given by:

$$\sum_{n=1}^{m} \binom{n}{9}$$

(4.1.9)

For e.g. m = 1, 2, 3 or 4, $\sum_{n=1}^{m} \binom{n}{9}$ = 9, 45, 129 or 255 masks respectively.

Clearly all n out of 9 possibilities must be specified in separate masks in the majority vote operation.This operation is clearly more based on the statistics of the neighbourhood than on the morphology that can be detected in the neighbourhood. This leads to the conclusion that another realization method based on counting pixels in a neighbourhood is necessary for this class of filters, rather than using an inexact matching approach. See also section 5.1.

The two most frequently used *recursive and iterative* cellular logic operations are: Connected component extraction or Propagation and Topology Preserving Thinning or Skeletonization.

- Connected component extraction or propagation, is a recursive neighbourhood operation between two binary images in which the pixels of one image (the seed) are propagated into the other image (the mask). The seed can also be formed by defining the image edge to be foreground; propagation starts then from the image edge inward. After propagation, all sets of connected pixels in the mask are selected which were also connected to the seed. The propagation operation belongs to the class of conditional dilations. It is also an iterative procedure. Figure 4.1.5 shows the structuring element for the propagation operation. Note that using the recursive neighbourhood speeds up the dilation.

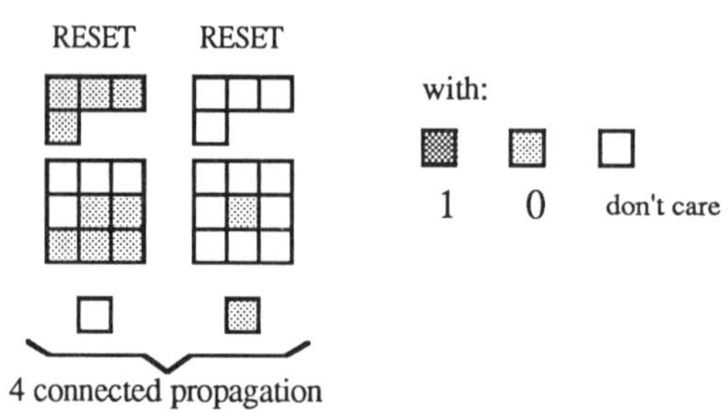

Figure 4.1.5 Mask set for the 4 connected propagation using the RESET definition (y_k is set to zero if one of the masks fits).

Moreover, when raster scanning is used to implement the transformation a notable speed-up is obtained if in all odd iterations the raster scan is performed from top-right to bottom left and in all even iterations the raster scan is performed from bottom right to top left. See figure 4.1.6.

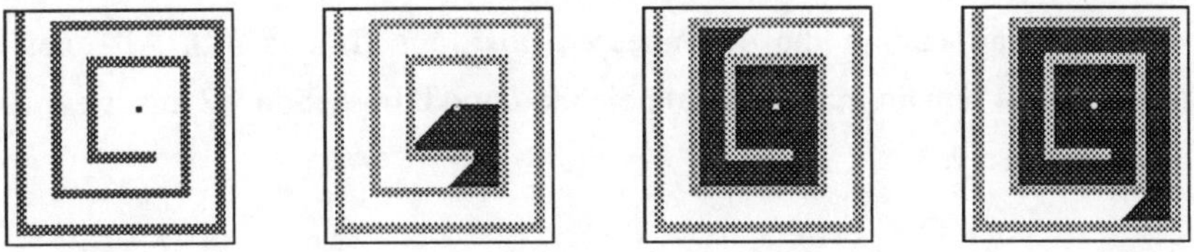

Figure 4.1.6 Propagation: Original and result after the first, second and third iteration over a spiral formed object in a binary image.

- Skeletonization, or topology preserving thinning or (in short) thinning, is informally defined here as a recursive neighbourhood operation to reduce objects to a one pixel thick line with the same topology as the original object (the skeleton). Preferably this line is the medial axis of the object. The skeletonization operation is a member of the class of conditional erosions and is usually applied iteratively until the image is stable. In section 4.2 a more formal approach to the thinning subject is given.
  Figure 4.1.7 depicts the mask set for the 8 connected skeleton. Note that it is one mask-set, but it can be envisioned to consist of four sub-sets.
  Subset a) is the erosion mask, it takes care of peeling the contour in each iteration and hence preserves the medial axis property of the skeleton.
  Subset b) consists of the connectivity conditions or breakpixel conditions that take care of the topology property of the skeleton.
  Subset c) are the endpixel conditions that take care of the sprouting of skeleton branches.
  Sub-set d) represents the single pixel condition; this mask prevents that in the worst case small objects can be eroded further than a single point; to void.

A number of observations on the skeletonization topic can be made at this point.

1) Subset b), the breakpixel masks, represent all possibilities for a skeleton curve to cross a 3 x 3 neighbourhood and are derived using the Hilditch crossing number approach (Hilditch 1969). The partial use of the recursive neighbourhood is necessary to prevent in each iteration the breaking of one pixel thick lines. Note that as in erosion, and hence in thinning, only the foreground (ones) can be set to background (zero) and never vice versa, only the zeroes need to be detected in the recursive neighbourhood. Moreover, because the foreground is 8 connected

and the background is 4 connected, the zeroes in the recursive neighbourhood will only appear on the 4 connected positions (N and W). All ones (the foreground) remain in the normal neighbourhood. In section 4.2 this is explained more precisely.

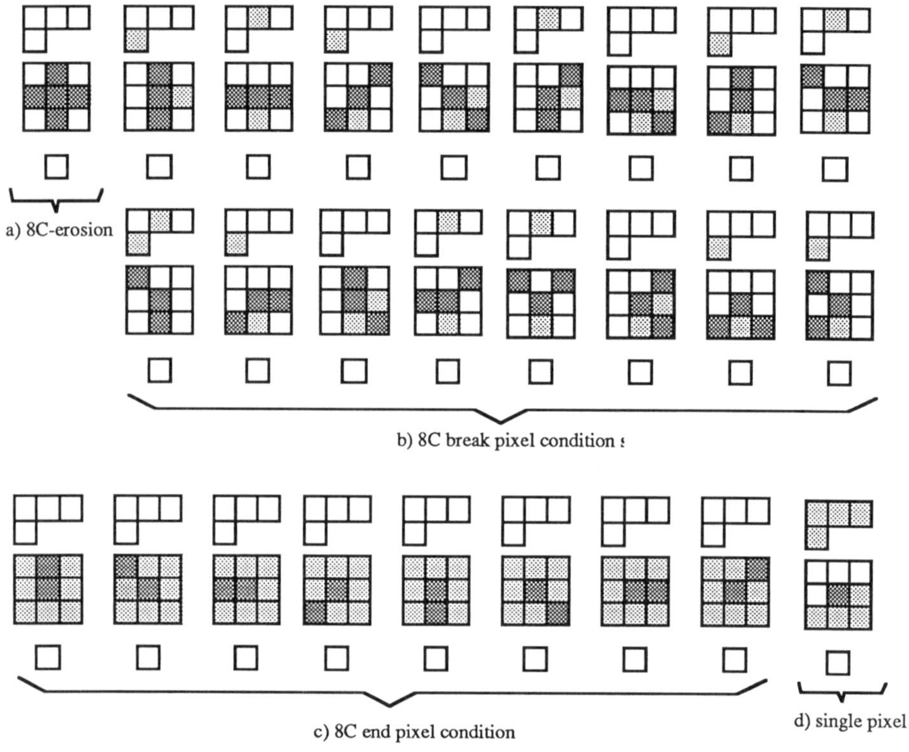

Figure 4.1.7 SET masks for the 8 connected skeleton
(y_k is set to one if one of the masks fits, else set to zero).

2) The thinning method presented here, is usually referred to as a sequential method, as it requires a sequential update of the pixels in the image. However, the neighbourhood match for each pixel can be performed in parallel. All masks can be tested in parallel or stored in a look-up table that can be used for a parallel match of all masks (see section 5.1).
Another thinning method is usually referred to as a parallel method as it uses a parallel update of the pixels in the image, as e.g. in Square Processor Arrays. By using the parallel update, usually *non-recursive* skeletonization schemes are used

(e.g. Arcelli et al. 1975). They are applied by iterating through the image using different masks in the *normal neighbourhood*, while each mask erodes in only one direction. This need to be done to prevent breakage of one pixel thick lines in the image. Consequently, processing these types of masks can only be performed in *a sequence* of eight inexact matches per iteration.

3) The way in which the neighbourhood is updated influences the final position of the skeleton. The final position of the skeleton points may be shifted one pixel in any direction. This depends on the direction from which the scan is performed. Note that this deviation is on top of the deviation caused by the erosion metric. The erosion mask as used in figure 4.1.7a performs the erosion of 8 connected contours; it implicitly uses a chessboard metric.
The decision where to place the final skeleton pixel when it should actually be placed in the middle of two remaining object pixels, depends on the update direction of the current iteration.
4) The SET specification of a mask set shows us which pixel configurations should remain in the resulting image after the transformation. This way of specification gives more insight in the result to be obtained than a RESET specification. This encourages the definition of new operations, which can be based on a specification of the required result after transformation.
5) Masks can be transformed to Look-Up Tables, by filling in the don't cares. Tables can be easily transformed from SET to RESET condition, by inverting the table value. By performing logic reduction (see e.g. Brayton et al. 1984b), tables can be reduced to mask sets. Through table comparison, transformation to SET conditions and logic reduction, the difference in behavior of any proposed skeleton scheme in literature and the Hilditch skeleton as reference can be listed.

4.1.4 Thinning variants.

Due to the fact that the mask-set of a skeleton can be split up into sub-sets that are in accordance with the required skeleton properties (erosion, breakpixels, endpixels, single pixel), skeleton variations can be easily made. This is elaborated in this subsection.

- A variant on the general skeleton scheme is the anchor-skeleton; A mask image (Z) is used, to force the skeleton through the points z_k specified in this mask image. The points z_k are inserted (logic OR) in the result image after each iteration. This skeleton can for instance be used to check the connectedness between a start and a goal point.

- The 4 connected skeleton. In section 4.2 the mask set of the 4-connected skeleton is derived. This skeleton has the nice property that its arcs are 4-connected which usually appeals more. However, its breakpixel mask set consists of 34 masks.
- The 4-connected skeleton with 8 connected erosion. This skeleton has a nice appearance due to the 4-connectedness of the remaining skeleton as well as a better skeleton metric. A combination of a 4 connected breakpixel mask set and an 8 connected erosion mask makes this possible. Moreover, the combination of this erosion mask with these breakpixel masks fortunately yields, after minimization, a very small breakpixel set of only 6 masks, as figure 4.1.8 reveals.
- Four connected skeletons use a cityblock metric for the erosion, consequently the arcs of the skeleton have preferences for north-south and east-west directions, biasing the skeleton. The eight connected skeletons use a chessboard metric for the erosion, resulting in, besides the NS and EW preferences, preferences for the diagonal directions. Skeletons with a near euclidian metric cannot easily be realized using only this mask approach. However, by using the 4 connected breakpixel set with -in each odd iteration- the 4 connected erosion mask and -in each even iteration- the 8 connected erosion mask, a slight improvement of the metric can be obtained[3].
- If the erosion is performed in a sequence dictated by the distance to the object border and the breakpixel and endpixel mask sets are used to preserve the topology, skeletons with near Euclidian metrics can be obtained. See (Verwer 1988) who used the (Borgefors 1986) Distance Transform with the DIP look-up tables (Gerritsen 1982) to generate a near Euclidian skeleton.
- It is possible to relax the skeleton properties by omitting subsets of the original mask sets. When omitting the endpixel conditions, objects with holes are eroded to closed contours, other objects are eroded to single points. Such a mask set is usable for blob-to-point reduction, if none of the objects in the image contains holes.

---

3 Note, that due to the larger erosion distance of the 4-connected erosion, this erosion cannot be combined with the 8-connected breakpixel set, but only with the 4-connected breakpixel set. This can be better understood after reading section 4.2, as it contains a forward reference to the term connectivity distance.

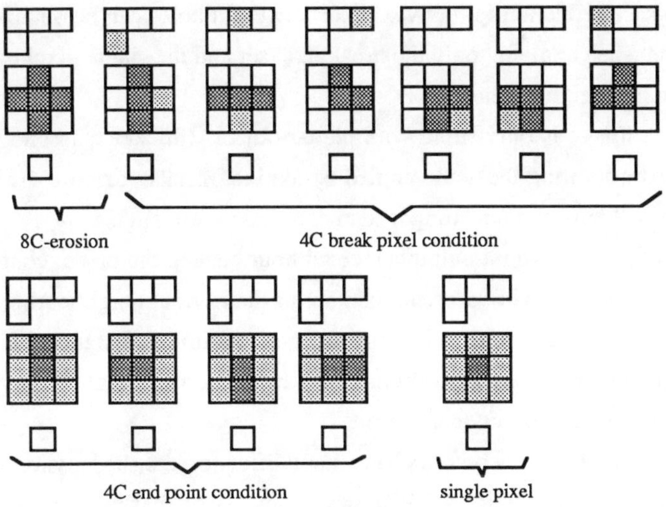

Figure 4.1.8 SET masks for the 4 connected skeleton with 8 connected erosion. (y_k is set to one if one of the masks fits, else set to zero).

- For any iterative cellular logic operation that uses the recursive neighborhood and a raster scan update technique, a faster performance is obtained if in all even iteration passes scanning takes place from top-left to bottom-right and vice versa in all odd passes. For skeletons without endpixel conditions the breakpixel set can be used in the full recursive neighbourhood. This enables a faster erosion, resulting in less iterations, especially when skeletonizing long thin objects. Long thin objects will erode to a single line in a few iterations. Normally the breakpixel conditions prevent fast erosion of these skeleton arcs; in each iteration only its endpoints are eroded. However, when also the ones in the breakpixel masks are applied in the recursive neighbourhood, in a downward scan all N-S, NW-SE, and W-E oriented skeleton ends are eroded -recursively- in a single downward iteration, and all S-N, SE-NW and E-W oriented skeleton ends are eroded -recursively- in one single upward iteration. (See also section 4.2)
- When skeletonizing images with noise, small objects can be removed if in the first passes the skeleton without endpixel conditions and without single pixel condition is applied.

- Blob-to-seed reduction for convex objects without holes can be obtained in a few iterations when applying only the breakpixel set and the single pixel mask both in the recursive neighbourhood.
- When the image is only filled with planar objects (thicker than one pixel in all directions), omitting the acute angled breakpixel masks of figure 4.1.7, the final skeleton will only contain straight segments and obtuse angles.
- A form of closing over a unit distance without biasing the object contours(!) can be obtained by applying a breakpixel set in the normal neighbourhood with the central pixel inverted. E.g. closing over a maximum distance of √2 can be obtained by using subset b) of figure 4.1.7 in the normal neighbourhood with the central pixel of the masks set to zero.
- Mostly, planing object contours has a good effect on the suppression of spurious skeleton branches. Figure 4.1.9 shows examples of masks that can be used to plane object contours.

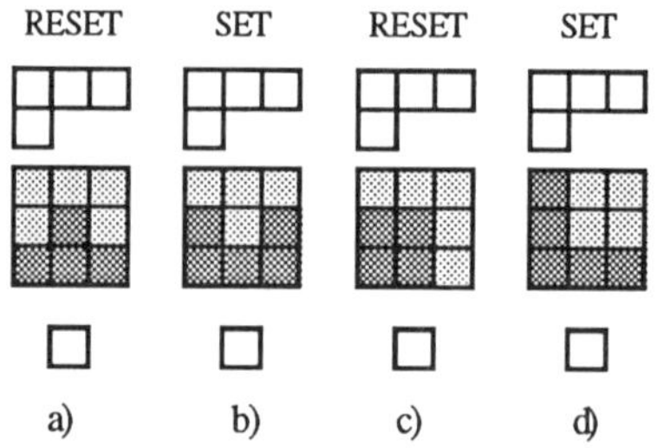

Figure 4.1.9 Masks to plane contours.

Mask a) and b) and their three rotation variants, plane contours in the horizontal and vertical directions, mask c) and d) and their three rotation variants, plane contours in the diagonal directions.The set masks and reset masks should be applied in two successive operations to avoid conflicts on battlemented contours.

If instead of the normal erosion condition in a skeleton mask set, in the first two iterations plane masks are taken, most spurious branches in the final skeleton are suppressed.

Figures 4.1.10 and 4.1.11 show some of the skeleton variants discussed.

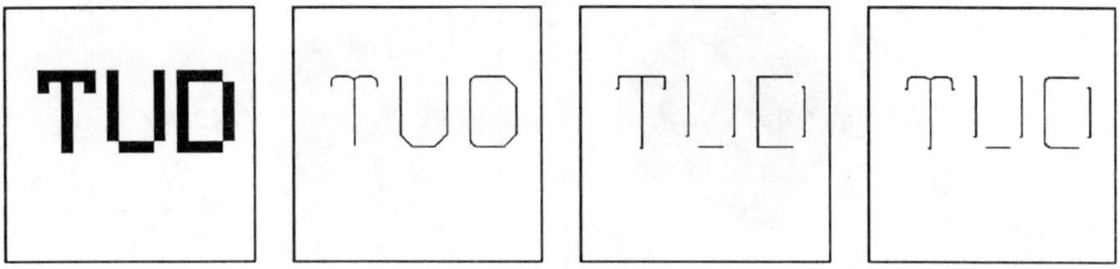

a) original b) 8c-skeleton c) 4c-skeleton d) 84c-skeleton

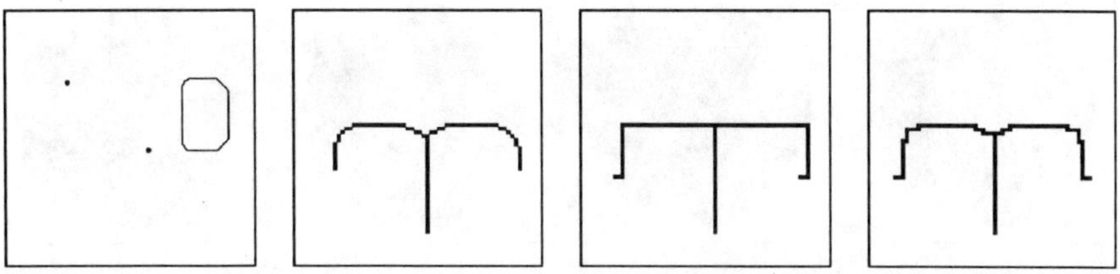

e) 8c-skeleton without endpixel condition f) detail of b) g) detail of c) h) detail of d)

Figure 4.1.10 Results of various skeleton variants.

Figure 4.1.10a shows the original image (named TUD). Figure 4.1.10b shows its 8-connected skeleton, using an erosion of 8-connected contours (Hilditch skeleton). Figure 4.1.10c shows its 4-connected skeleton using an erosion of 4-connected contours. Figure 4.1.10d shows its 4-connected skeleton using an erosion of 8-connected contours. Figure 4.1.10e shows its 8-connected skeleton using an erosion of 8-connected contours without endpixel condition, but with single point condition. Note, that the resulting points of the letters T and U are dilated to improve the visibility for the reader of this book. The skeleton conditions are used in the full recursive neighbourhood (see also section 4.2.12). The figure 4.1.10f, g and h show a zoomed version of the skeletons of the figures 4.1.10b, c and d.

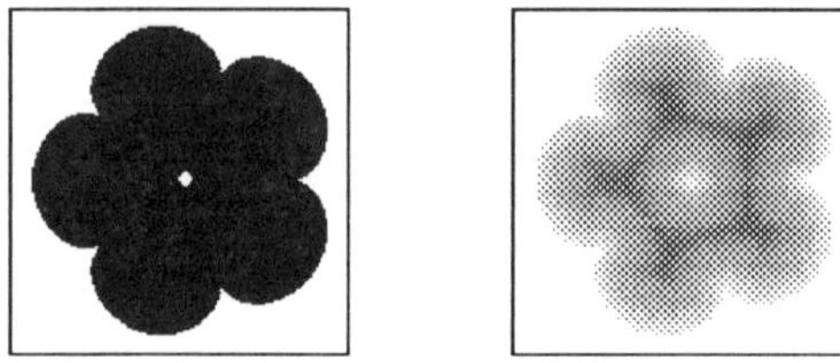

a) Original image (flower) b) (5,7) distance transformed image

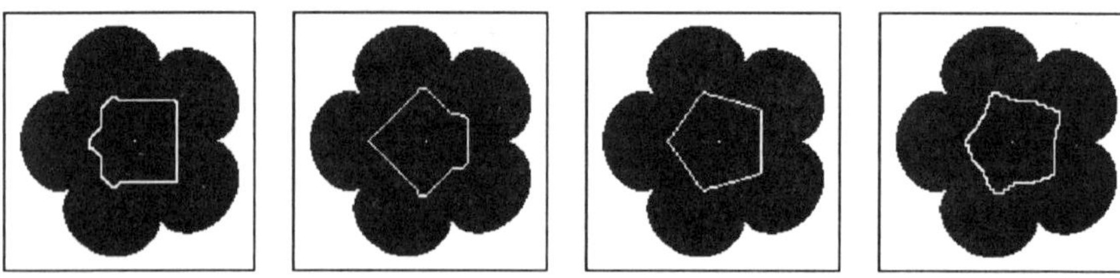

c) 4c-skeleton d) 8c-skeleton e) (57dt)8c-skeleton f) (4/8)4c-skeleton

Figure 4.1.11 Results of various skeleton variants to show the influence of the erosion metric.

Figure 4.1.11a shows the original image with a single background pixel in its center (named Flower). Figure 4.1.11b shows the (5,7) distance transform of Flower (57dt). Figure 4.1.11c shows the 4-connected skeleton superimposed on Flower. Figure 4.1.11d shows the 8-connected skeleton superimposed on Flower. Figure 4.1.11e shows the 8-connected skeleton based on erosion in order of the distances to the border and an 8-connected breakpixel set. Figure 4.1.11f shows the 4-connected skeleton without endpixel condition, however by repeatedly interchanging the erosion metric from 4 to 8 and vice versa. The metric shows some similarity with the distance transform skeleton, however, it is entirely based on cellular logic (bit) operations, in contrast with the distance transform that is based on integer arithmetic.

4.1.5 Conclusions.

In this sub-section an informal and practical approach to Cellular Logic Processing in two dimensional images has been presented. The method as developed, appeared to be advantageous in many ways. It is based on mathematical morphology and, moreover, it visualizes the morphological properties of the transformation. In this way it became possible to predict the result of the transformation by looking at the mask set of the transformation. This set-up facilitates making variations on known transformations, comparing the effect of known transformations of different authors with each other, and proposing new transformations. New transformation schemes are obtained by specifying -in a mask set- all the situations that can locally be found - in a 3 x 3 neighbourhood- anywhere in the output image.

The extension of this mask set approach to cellular logic operations in high dimensional images is treated in the next section.

The development of special hardware to execute in parallel, in a single clock-cycle the match of the neighbourhood with the mask-set, is treated in chapters 5 and 6.

4.2 Conditions for multi-dimensional thinning.

4.2.1 Introduction.

In this section we will provide insight into the elements that are necessary to constitute a skeleton in N dimensions. Using these elements, we will produce a method for the construction of N dimensional thinning tools, here applied for N≤3. Apart from generating thinning conditions, the described method can be used to get insight into the various existing thinning schemes and be applied to compare their performances.

A short review on topology preserving thinning or skeletonization is presented first. Thinning or skeletonization in 2-D has received much attention in literature. Recent overviews are given by (Lee et al 1991) and (Lam et al 1991). Two classical 2-D skeletonization schemes are:

On the one hand, the parallel image update, sequential neighbourhood match approach, as e.g. proposed by (Arcelli et al 1975) in the form of a set of masks (3 x 3) for a massively parallel processor such as the CLIP-4 (Duff 1982).

On the other hand, the sequential image update, parallel neighbourhood match approach, commonly used in software, based on a crossing number as was proposed by (Hilditch 1969). The parallel neighbourhood match in the sequential approach is often performed using a look-up table.

For 3-D skeletonization some of the oldest references lead to (Lobregt et al 1980) and (Toriwaki et al 1982), Lobregt using an extended Euler number count procedure to create look-up tables for 2 x 2 x 2 neighbourhoods. With a sequence of table look-ups the value of the central pixel was obtained.

To obtain a near-Euclidean skeleton first in 2-D, (Verwer 1988) combined a 3 x 3 table look-up approach for the connectivity test with a (5-7-11) distance transform (Borgefors 1986) and later in 3-D, combined the 2 x 2 x 2 table sequence of Lobregt extended with situations not detectable on a square grid by Euler count, as was remarked by Toriwaki, with a 3-D distance transform. Verwer obtained a fast implementation by using a bucket sorting queue to store, in order of their distance to the border, only the pixels or voxels that were likely to be eroded in the next iteration (Verwer 1991).

A recent work of (Preston 1991) describes 3-D skeletonization on a 3-D equivalent of a hexagonal grid using an Euler number count procedure. The neighbourhood match was performed using a table look-up (size 2^{13}), incorporated in a special hardware architecture. The 3D-grid used, with a central pixel and 12 neighbours, has the advantage that table look-up is possible even in 3-D. However, resampling in order to transform the image originally acquired from a square grid to the new grid is often inevitable and may introduce undesirable distortions.

A short *abstract* of the contents of this chapter is given in the next paragraphs.
For thinning on a square grid, with large neighbourhoods or high dimensions, a straightforward look-up table approach does not seem suitable any more, due to the large storage space required to store each possible neighbourhood configuration ($\geq 2^{26}$). Consequently, it is highly unlikely that a special architecture for skeletonization on a square grid in higher dimensions can be based on a straightforward table look-up. As skeletonization can be easily described with sets of structuring elements (sets of templates or sets of masks) and as these sets can be sparsely stored in hardware (Jonker and Duin 1985a, Jonker et al 1988b) descriptions with sets of 3 x 3 binary structuring elements may fulfill both hard- and software requirements. This method, extended to N-dimensional images, is the subject of this chapter.
In the following sub-sections first the definitions of an N dimensional binary -image, -image edge, -image element, -structuring element and -neighbourhood are given.
Next the commonly used terms for the connectivity between image elements is extended with a definition of the connectivity of elements in an N-dimensional image in terms of the Euclidean distance between those elements. This is done for two reasons: firstly, to provide a consistent description of the connectivity for N>3, and secondly, to define the layer thickness of object boundaries, which is discussed later on in this chapter.
The discourse is continued with some definitions of basic objects in an N-dimensional image and their intrinsic dimensionality $\tilde{N}$, such as curves and planes in 2-D, space curves, curved surfaces and volumes in 3-D. The basic objects are defined such, that all objects in an image can be described as compound structures composed of these basic objects.
Next, the most suitable connectivity for the foreground and background of an N-dimensional binary image is discussed, based on the permeability of the boundary of an object. The most suitable connectivity for objects with a specific intrinsic dimension is also discussed, based on the desired curvature of the basic objects.
The basic objects are described in terms of piecewise polynomials and the transformation between the geometrical domain and the image element domain – and

its inverse – is established. This is done for two reasons. Firstly it gives a better mathematical background to the maximum curvature of basic objects, and secondly the geometrical description can be used to determine the intersection of objects.

Then, sets of binary structuring elements or sets of masks are defined as being constituted from an intersection of foreground and background object primitives. An object primitive is the smallest possible object that still has the property of a specific basic object, analogous to molecules in chemistry. Each object primitive in turn consists of one or more tiles, analogous to atoms in chemistry. The mask sets are able to detect basic objects, for example a set that hits on (matches with) all surfaces in 3-D. Furnished with a suitable number of "don't cares", the structuring elements are also able to match compound objects in an image, e.g. ones with forking curves, forking surfaces, curves emerging from volumes etc.

Which foreground and background object primitives are intersected depends on the connectivity and permeability of the basic objects they represent.

Finally, N-dimensional skeletonization can then be defined as a conditional erosion using sets of masks. These sets consist of:

a) A set specifying the erosion-condition and erosion metric for dimension N.
b) Sets specifying the break point conditions that prevent breaking of the topology.
c) Sets specifying the sprouting of skeleton ends.

The break point set matches on all compound objects in the image composed of basic objects with an intrinsic dimension $\tilde{N} \leq N - 1$ and thus preventing them from being eroded. For example, in 3-D, on all objects composed of space-surfaces and space-curves (and single point objects).

In the Appendices mask sets for various skeletons in 2-D and 3-D are presented.

4.2.2 Definition of an N dimensional binary image, image edge and image element.

If $\mathbb{R}_N$ is a Euclidean space of dimension N with origin $\vec{O}$ then let $\mathbb{R}_N^{\cdot}$ be a Euclidean space of dimension N with origin $\vec{O}$ equidistantly sampled with unit distance over each dimension. An N-dimensional binary image X_N is now defined as an N-dimensional bounded section of $\mathbb{R}_N^{\cdot}$, with the elements of X_N having the values $\{0,1\}$. See figure 4.2.1a. The size of the image is indicated by the vector $\vec{s}_N$: $(s_1,..,s_N)$ containing the bounds of each coordinate. Let X_N' be an image of size $\vec{s'}_N$: $(s_1+2,..,s_N+2)$ having its origin in $\vec{O} - \vec{1}$, then the edge ε_N of X_N is defined as the elements of $X_N' \notin X_N$. The elements of X_N' and ε_N also have the values {0,1}. The elements of X_2 are referred to as pixels, the elements of X_3 as voxels.

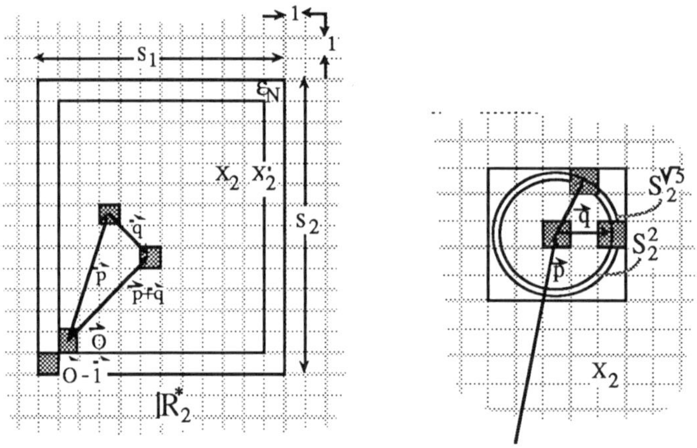

Figure 4.2.1 a) Binary Image X_2 in $\mathbb{R}_2^*$. b) Neighbourhood M_2^5 of $\vec{p}$.

4.2.3 Connectivity of image elements and neighbourhood in an N dimensional image.

Let the position of an element in an N-dimensional image X_N be denoted as $\vec{p}$, then $\vec{p}+\vec{q}$ denotes an element with position $\vec{q}$ relative to $\vec{p}$. See figure 4.2.1b. We now make the following definition, presupposing that each vector points to the center of an image element:

> **Definition**: *All elements of* X_N *with a distance* $d \leq |\vec{q}|$ *to* $\vec{p}$ *are a set of elements connected to* $\vec{p}$ *and are assumed to lay within the (hyper-) sphere* S_N^d *with origin* $\vec{p}$ *and radius* d.

Let M_N^n be an N-dimensional (hyper-)cubic neighbourhood with (odd) size $n = 2k+1$, having its central element in $\vec{p}$. If $\vec{p}$ is assumed to be the origin of the local coordinate system, then k is the maximum value of any component of $\vec{q}$ within M_N^n.

Let E be the number of elements **on** the (hyper-)sphere S_N^d **within** the neighbourhood M_N^n. As the elements are exactly on the grid positions, E will only have non-zero values for specific values of *d*.

In order to derive an expression for E(N,k,d), let us consider only the elements $\vec{q}$ in the **partition** with non-negative component values, i.e., with $0 \leq q_i \leq k$ within the neighbourhood M_N^n. For N=2, this means considering only elements in the first **quadrant**, for N=3 only in the first **octant**, etcetera. Afterwards the number found for such a partition can be multiplied by the number of partitions, 2^N and compensated for the shared partition boundaries.
The number of different vectors $\vec{q}$ with the same length $|\vec{q}|$ is equal to the number of **permutations** among its component values. So, if all N components of $\vec{q}$ have different values, there will be N! vectors with length $|\vec{q}|$ in the partition.
Let us denote the number of times that each of the components q_i of $\vec{q}$ has the value j by n_j, i.e.:

$$n_j = \sum_{i=1}^{N} (q_i = j)$$

(4.2.1a)

Then, for each distinct $0 \leq j \leq k$, the number of vectors with length $|\vec{q}|$ will be reduced by a factor $(n_j!)^{-1}$, because permutations of equal component values do not produce different vectors. Note that if only two component values are possible, e.g. 0 and 1, the result is the binomial:

$$\frac{N!}{n_j!(N-n_j)!}$$

(4.2.1b)

The more general case, with more than two different component values is called **multinomial:**

$$\frac{N!}{\prod_{j=0}^{k} (n_j!)} \quad \text{with} \sum_{j=0}^{k} n_j = N$$

(4.2.1c)

The case j = 0 is a special one, because a vector with one or more components $q_i = 0$, i.e. with $n_0 > 0$, is shared by 2^{n_0} partitions (quadrants, octants, etcetera). So, instead of simply multiplying afterwards by the number of partitions 2^N, we must use the factor $2^{N - n_0}$ in order to compensate for shared vectors. So, in conclusion E(N,k,d) is given (for $d \leq k$) by:

$$E = \frac{N!}{\prod_{j=0}^{k} (n_j!)} 2^{(N-n_0)}$$

(4.2.1d)

By way of example, figure 4.2.2 shows a positive quadrant in X_2, in which there are two (edge-edge connected) elements lying on the circle with $d = 2$: $\vec{q} = (0, 2)$ and $\vec{q} = (2, 0)$. According to (4.2.1c), their number is indeed $\frac{N!}{n_0! \cdot n_1! \cdot n_2!} = \frac{2!}{1! \cdot 0! \cdot 1!} = 2$.
According to (4.2.1d) the number of elements lying on the **complete** circle with $d = 2$ is not four (the number of quadrants) times as many, but only twice, because both elements in the first quadrant are shared with another quadrant.
Likewise, the number of (point-edge or knight's move connected) elements in the first quadrant lying on the circle with $d = \sqrt{5}$ is: $\frac{2!}{0! \cdot 1! \cdot 1!} = 2$. Because in this case none of the elements is shared among partitions (quadrants), their number on the full circle is 8, simply 2^2 (the number of quadrants) times as many.
In 3 dimensions the number of (knight's move connected) elements on a sphere with $d = 3$ in the positive octant is: $\frac{N!}{n_0! \cdot n_1! \cdot n_2!} = \frac{3!}{0! \cdot 1! \cdot 2!} = 3$, i.e. $\vec{q} = (1, 2, 2)$, $\vec{q} = (2, 1, 2)$ and $\vec{q} = (2, 2, 1)$. Because again all the vector components are non-zero, the number of elements on the **complete** sphere with $d = 3$ will be 24, 2^3 (the number of octants) times as large.

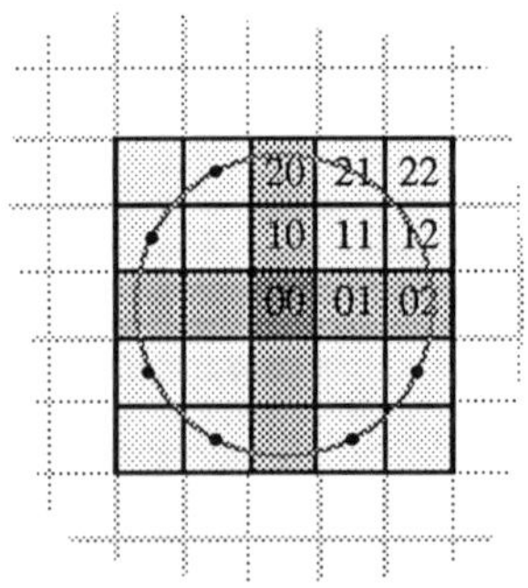

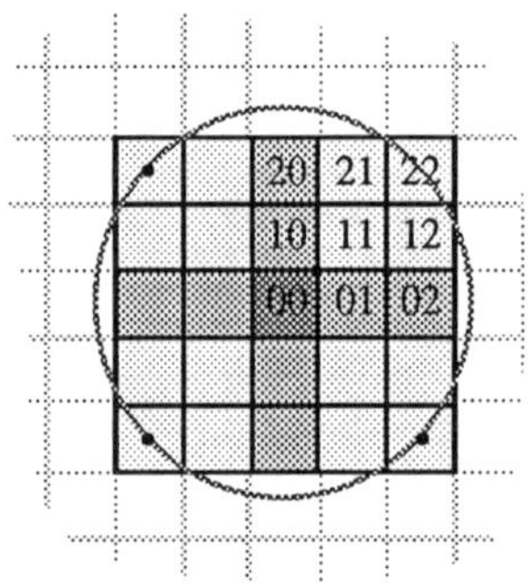

Figure 4.2.2 The vectors in the positive quadrant in X_2 for k = 2.

Table 4.2.1 shows E, the number of elements within M_N^n and **on** S_N^d for some dimensions and neighbourhood sizes, V, the number of elements within M_N^n and

within S_N^d, and G = V–1, commonly used to indicate the connectivity between elements. We will refer to the sphere radius *d* as the **connectivity distance *d*.** Note that this distance is used in sub-section 4.2.5 to define the thickness and permeability of foreground and background objects.

Neighbourhood size: M_N^n =	Connectivity type:	Typical $\vec{q}$ =	Sphere radius: *d*=	Elements **on** sphere: E =	Elements **in** sphere: V =	Neighbourhood connectivity: G =
1	edge	[0]	√0 = 0	1•1 = 1	1	0
3	point	[1]	√1 = 1	1•2 = 2	3	**2**
5	point-point	[2]	√4 = 2	1•2 = 2	5	4
1x1	face	(0,0)	√0 = 0	1•1 = 1	1	0
3x3	edge	(0,1)	√1 = 1	2•2 = 4	5	**4**
	point	(1,1)	√2 = 1.4	1•4 = 4	9	**8**
5x5	edge-edge	(0,2)	√4 = 2	2•2 = 4	13	12
	point-edge	(1,2)	√5 = 2.2	2•4 = 8	21	20
	point-point	(2,2)	√8 = 2.8	1•4 = 4	25	24
1x1x1	volume	(0,0,0)	√0 = 0	1•1 = 1	1	0
3x3x3	face	(0,0,1)	√1 = 1	3•2 = 6	7	**6**
	edge	(0,1,1)	√2 = 1.4	3•4 =12	19	**18**
	point	(1,1,1)	√3 = 1.7	1•8 = 8	27	**26**
5x5x5	face-face	(0,0,2)	√4 = 2	3•2 = 6	33	32
	edge-face	(0,1,2)	√5 = 2.2	6•4 = 24	57	56
	point-face	(1,1,2)	√6 = 2.5	3•8 = 24	81	80
	edge-edge	(0,2,2)	√8 = 2.8	3•4 = 12	93	92
	point-edge	(1,2,2)	√9 = 3	3•8 = 24	117	116
	point-point	(2,2,2)	√12= 3.4	1•8 = 8	125	124
1x1x1x1	hypervolume	(0,0,0,0)	√0 = 0	1•1 = 1	1	0
3x3x3x3	volume	(0,0,0,1)	√1 = 1	4•2 = 8	9	**8**
	face	(0,0,1,1)	√2 = 1.4	6•4 = 24	33	**32**
	edge	(0,1,1,1)	√3 = 1.7	4•8 = 32	65	**64**
	point	(1,1,1,1)	√4 = 2	1•16 = 16	81	**80**

Table 4.2.1 Neighbourhood connectivity as a function of dimension, neighbourhood size and distance.

Unfortunately, for larger values of k and N, even within one partition more than one set of values n_j, j=0...k (cf. 4.2.1a), that results in the same d, and E(N,k,d), may occur. For N=2, this occurs with k≥5, because, e.g., |0,5| = |3,4|. Likewise, for N=3, with k≥3, e.g. |0,0,3| = |1,2,2|. For N≥4, it occurs in all neighbourhoods with k≥2: e.g. |0,0,0,2| = |1,1,1,1|, |0,0,0,0,2| = |0,1,1,1,1|, etc. This ambiguity for d does **not** exist for k=1, because in this case $d=\sqrt{n_1}$, which is different for every distinct n_1-n_0-combination, and no other combinations than these (of vector components with values of either 0 or 1) are possible. To avoid the ambiguity, we will restrict ourselves to k=1.

Whereas E is only non-zero for **distinct** values of d (delta function), G is non-zero for **continuous** values of d (step function). Like E, G is a function of N, k and d. From this point onward we will restrict ourselves to the connectivity defined in neighbourhoods with k = 1, to avoid the ambiguity discussed in the preceding paragraph. Note that there is no **fundamental** objection against the concept of connectivity defined in larger neighbourhoods, such as with k = 2, when a knight's move is allowed to connect pixels. It is just the **description** using only N, K and d, that is no longer unambiguous with N≥4.

Applying the restriction that k = 1, i.e., allowing only 3^N neighbourhoods, the connectivity between two **image elements** will be denoted by G_N^d. (Note that with this restriction the multinomial distribution degenerates to a binomial distribution.) Consequently, we will make the following

Definition: *An element* $\vec{p}+\vec{q}$ *is said to be* G_N^d*-connected to* $\vec{p}$ *if* $|\vec{q}| \leq d$ *and* $max\,(q_i) = 1$.

For example the element (1, 1, 1) is said to be $G_3^{\sqrt{3}}$-connected to (0, 0, 0). Throughout this article we will also use the more common notations from table 4.2.1, such as point-connected or 26-connected.

The figures 4.2.3a and 4.2.3b show the neighbourhoods M_2^3 and M_3^3 with their connectivity possibilities G_2^1, $G_2^{\sqrt{2}}$ and G_3^1, $G_3^{\sqrt{2}}$, $G_3^{\sqrt{3}}$.

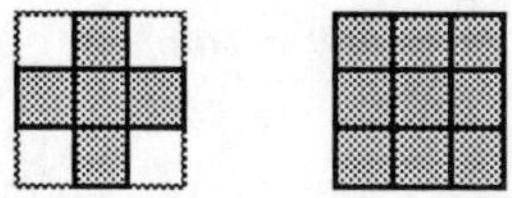

Figure 4.2.3 a) 4- and 8-connected elements (pixels) in neighbourhood M_2^3.

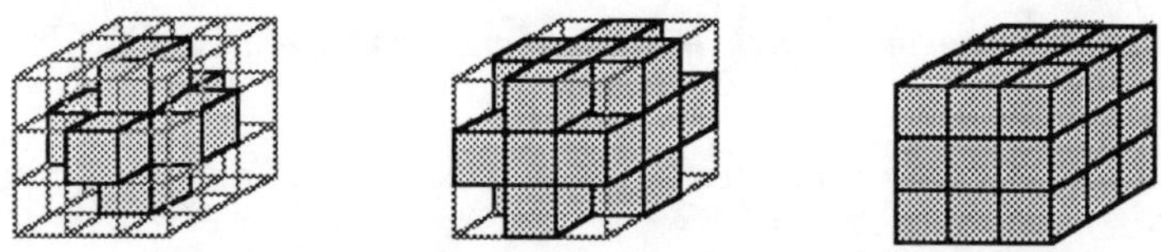

Figure 4.2.3 b) 6-, 18- and 26-connected elements (voxels) in M_3^3.

4.2.4 The structure of objects in N dimensional images.

In the preceding sub-section a definition of connectivity between the elements of an N-dimensional image was given. In this sub-section objects in an image are stepwise refined, describing them as being composed of smaller, simpler structures.
The motivation for this approach is, that in the thinning process objects are reduced to simpler structures. Planar structures are reduced to curved structures in X_2, voluminous structures are reduced to surface structures in X_3. In this sub-section an ordering scheme of all object structures in an image is established.

Definition: *An object in a binary image* X_N *is defined as a set of mutually connected image elements* $\vec{p}$ *with the same value.*

Note that an image may contain more than one object. If an image has more than one object all objects have the same value. For simplicity reasons we will assume the value 1 for all image elements of all objects in X_N.

Definition: *The set of all objects in an image is called the image foreground.*

Definition: *The set of all non-object image elements in an image is called the image background. The background of image* X_N *has the inverse value of the foreground.*

Hence it has the value 0 in our case.

Proposition: *Each object in an image is completely enclosed by background.*

Definition: *The connectivity* G_N^d *of an object is the largest connectivity of any two constituting image elements of the object.*

Definition: *A basic object* $O_{N,\tilde{N}}^d$ *is a* *<u>non-forking</u>* *object with any arbitrary shape and connectivity, having a single intrinsic dimension* $\tilde{N}$*, with:* $0 \leq \tilde{N} \leq N$.

The intrinsic object dimensions can be categorized for the first three dimensions as follows:

N = 1

- $\tilde{N} = 0$: Point
- $\tilde{N} = 1$: Line

N = 2

- $\tilde{N} = 0$: Point
- $\tilde{N} = 1$: Line or curve
- $\tilde{N} = 2$: Plane

N = 3

- $\tilde{N} = 0$: Point
- $\tilde{N} = 1$: Line or space curve
- $\tilde{N} = 2$: Planar or curved surface
- $\tilde{N} = 3$: Volume

For example $O_{3,2}^{\sqrt{2}}$ is a non-forking 18-connected (possibly curved) surface in X_3. $O_{3,1}^{\sqrt{3}}$ is a non-forking 26-connected line or space curve in X_3. $O_{2,1}^{1}$ is a non-forking 4-connected line or curve in X_2. $O_{2,2}^{1}$ is a 4-connected plane in X_2. Note that forking and curving is only possible when $0 < \tilde{N} < N$. For $\tilde{N} = 0$ the basic object's size is the unit size, which prevents forking or curving. For $\tilde{N} = N$ the basic object's structure spatially fills all dimensions and leaves no freedom for forking or curving. $N - \tilde{N}$ are the degrees of freedom for the basic object.

Definition: *A compound object is an object of any arbitrary shape and size composed of one or more basic objects .*

This definition covers all objects in image X_N. It leads for example in X_3 to objects composed of a number of volumes and / or surfaces and / or space curves and in X_2

to objects composed of a number of planes and / or curves. Objects with a size $\leq M_N^3$ are special cases and are treated below.

Definition: *An object primitive* $P^d_{N,\tilde{N}}$, *is a basic object of size* M^3_{N}.

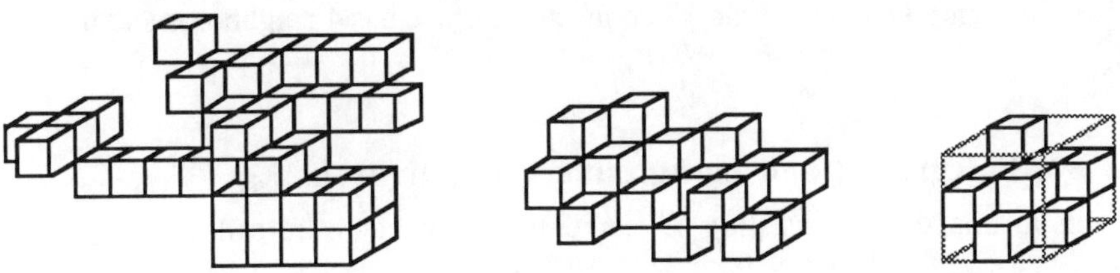

Figure 4.2.4 A compound object, a basic object (surface) and an object primitive.

Definition: *A tile* $T^d_{N,\tilde{N}}$ *is a basic object of size* $M^2_{\tilde{N}}$.

Due to the fact that the object primitive is defined in the neighbourhood M^3_N, curvature can always be modeled. As the tile is only defined in the M^2_N neighbourhood, only non-curved (flat) and non-forking basic objects can be modeled.

Figure 4.2.5 a) line tiles in X_2 b) plane tiles in X_2.

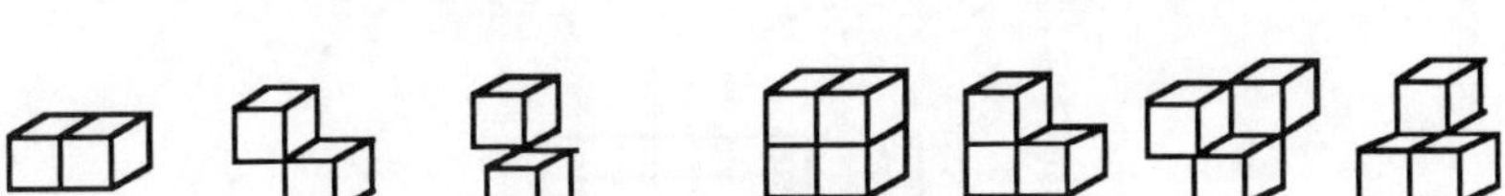

Figure 4.2.6 a) line tiles in X_3 b) plane tiles in X_3.

Figure 4.2.7 Volume tiles in X_3.

Note that an object primitive $P^d_{N,\tilde{N}}$ can be tessellated from tiles $T^d_{N,\tilde{N}}$ and that all possible variants in between the object primitive and the tile exist. We will call these variants partial object primitives ($Pp^d_{N,\tilde{N}}$).

Note that this is a projection: A hypersphere projects to a circle, e.g. $G_3^{\sqrt{2}}$ projects to $G_2^{\sqrt{2}}$, or the central element of an 18-connected surface has 4 neighbours and hence 4 tiles.

4.2.5 Foreground and background connectivity in X_N.

Objects in image X_N consisting of elements with value 1 are called **foreground objects.** The foreground objects are compound structures composed of basic objects with their own intrinsic dimension. The basic objects in turn are envisioned to be composed of object primitives, the smallest possible basic object. Due to the M_N^3 neighbourhood, object primitives may still be curved, whereas tiles, the building bricks of object primitives are always flat as they are defined in a M_N^2 neighbourhood. Like the foreground, the background can be envisioned as being constituted from objects. They will be addressed as **background objects** and the formulation of compound objects, basic objects, object primitives and tiles can be utilized on the background too.

As a consequence of this model of foreground and background objects, foreground objects may be enclosed by background objects and vice versa. Figure 4.2.8 shows a 2-D case in which two foreground objects are separated by a compound background object, and one background object is enclosed by a compound foreground object.

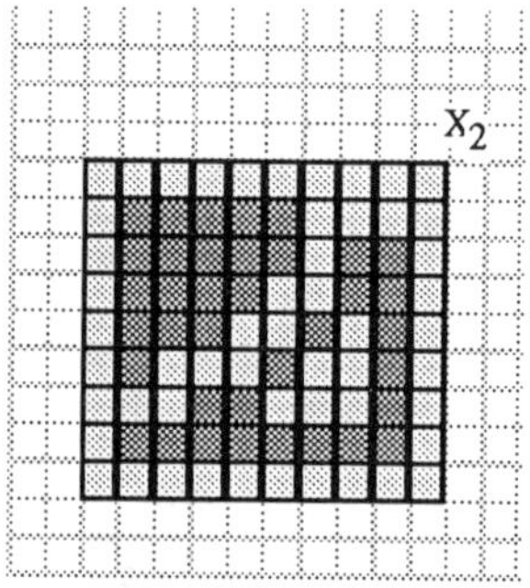

Figure 4.2.8 Model of objects and background in image X_2.

The compound background object is formed by a plane and three closed curves. The compound foreground object consists of two planes connected by two curves and one small line segment sprouting from its lower left plane.

The enclosing background's task is to separate objects. Consequently the background enclosure should be thick enough to prevent the touching of the foreground objects. This thickness depends on the connectivity and the smallest objects to measure this thickness on are the tiles. Figure 4.2.9 shows how a thickness measurement is set up.

Suppose a tile to be perforated (probed) consists of small (hyper)spheres or cannonballs[1] all laying on surface-to-surface distance *a* from each other. Another (perforating) tile consists of big spheres all laying on distance *b* from each other. It is presumed that the center lines of perforating and probed tiles are intersecting each other between or in the centers of the spheres. From this point onward we will refer to the segment of the center line in between and including the center points as the center line of the tile.

The probed tile (small spheres) is supposed to separate the perforating tile (big spheres) if $a < b$ and the perforating tile is supposed to perforate the probed tile if $a > b$. If $a = b$ no judgment can be given. If either one of the distances is zero the spheres of the corresponding tile make surface contact.

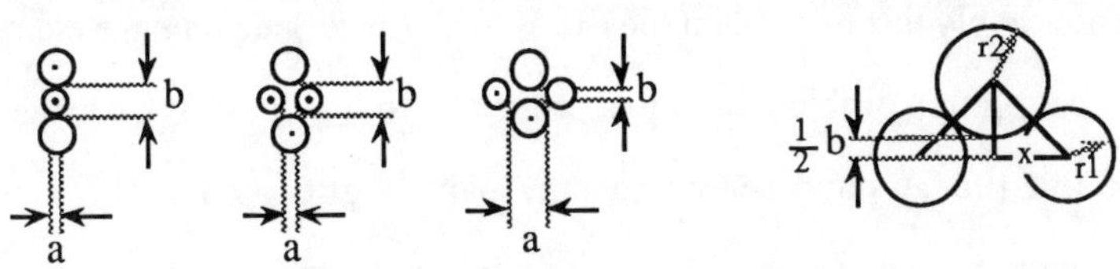

Figure 4.2.9 Measurement model.

[1] The terminology used in the field of crystallography is somewhat different (Ashcroft and Merin 1976, Kittel 1986). For the term grid, the term Bravais lattice is used: "An infinite array of discrete points with an arrangement and orientation that appears exactly the same, from whichever of the points the array is viewed." The M_N^2 neighborhood would be depicted as a primitive unit cell. The M_N^3 neighborhood would be depicted as a face centered cubic lattice (fcc). A unit cell of the sodium chloride crystal is an example of a "26 connected" structure on a fcc lattice.

If the radii of the small and big sphere are called r_1 and r_2 respectively, then the surface-to-surface distance b of the big spheres as a function of the distance of the small spheres is given in the 2-D case by (see figure 4.2.9):

$$b = 2\{ \sqrt{(r_1+r_2)^2 - x^2} - r_2 \} \tag{4.2.2a}$$

If $x = 0$ then $b = 2r_1$ and if $x = \sqrt{r_1^2 + 2r_1r_2}$, then $b = 0$, the first and third situations of figure 4.2.9. It will be clear that the maximum possible distance between the spheres of the perforating tile is obtained when their center line is in line with the center of one of the spheres of the probed tile. The minimum possible distance is always obtained if the spheres of the perforating tile are halfway the two spheres of the probed tile, independent from the length of either center lines.

Definition: *The layer thickness* D_N^d *of a probed tile is the length of the center line of the perforating tile.*

With the radii of all spheres equal (to r) this yields in X_2:

$$D_2^d = 2\sqrt{4r^2 - x^2} \tag{4.2.2b}$$

We will now apply this method on tiles $T_{N,\tilde{N}}^d$ with (hyper)cubic image elements. First bring the perforation problem back to a probe of two line tiles $T_{N,1}^d$. Hence, consider a probed line tile $T_{N,1}^d$ and search for any pair of grid points which center line intersects with the center line of $T_{N,1}^d$ with minimal length. The tile formed by these grid points will be designated a perforating tile. See figure 4.2.10.

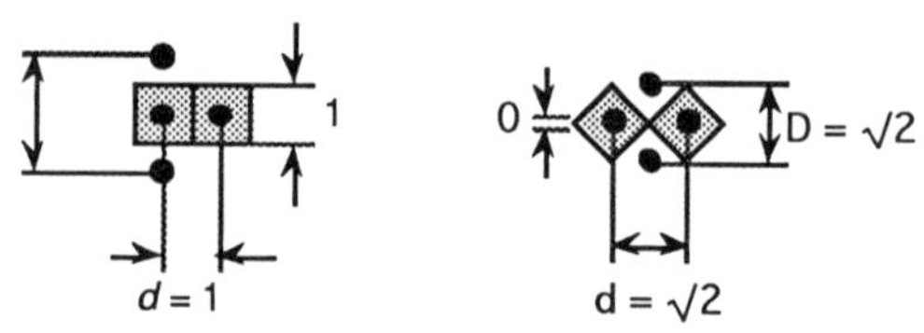

Figure 4.2.10 Layer thickness and connectivity distance in X_2.

In contrast with spheres, the radii of the cubic image elements and hence the layer thickness depends on the orientation of the image elements in $T^{d}_{N,1}$.

For example: If $x = 0$ and $r = \frac{1}{2}d = \frac{1}{2}$ then $D^{d}_{2} = 2$.

If $x = \frac{1}{2}d = \frac{1}{2}\sqrt{2}$ and $r = \frac{1}{4}d\sqrt{2} = \frac{1}{2}$ then $D^{d}_{2} = \sqrt{2}$.

It can be easily seen now that the layer thickness D^{1}_{N} is always ≥ 2 due to the fact that on all possible minimal distance grid positions, the center line of the perforating tile and the center of one of the image elements of the probed tile are in line ($x = 0$).
The layer thickness $\{D^{d}_{N} \mid (d > 1)\}$ is always d, the connectivity distance of the probed tile, due to the fact that on any grid position the center line of the perforating tile and the center of the probed tile are in not in line but intersect halfway ($x = \frac{1}{2}d$), where the surface-to-surface distance is zero [2].
For all other probed tiles $T^{d}_{N,\tilde{N}}$, i.e. tiles consisting of more than two image elements, possibly with different connectivities, the measurement has to be performed on all image element pairs of the tile. The resulting layer thickness is the minimum of all measured distances.

Definition: *A* G^{d}_{N}*-connected tile can perforate a tile* $T^{d}_{N,\tilde{N}}$ *with layer thickness* D^{d}_{N}*, if the connectivity distance d of the* G^{d}_{N}*-connected tile is larger than, or equal to, the layer thickness* D^{d}_{N} *of* $T^{d}_{N,\tilde{N}}$.

This leads to the conclusion that G^{1}_{N}-connected objects are the only objects that cannot be perforated by, nor can perforate any other G^{d}_{N}-connected object and that all other $\{G^{d}_{N} \mid (d > 1)\}$-connected objects can perforate any other $\{G^{d}_{N} \mid (d > 1)\}$-connected object.

Consequently:

---

[2] Note that these connectivity problems do not exist in images sampled on a hexagonal grid or equivalent in higher dimensions. On the intersection point half way the surface-to-surface distance is minimal but never zero.

Observation: *A proper choice for the background connectivity in image X_N is the G_N^1-connectivity, as it prevents leakage of foreground via background and, moreover, it is not able to perforate the foreground.*

The latter property is useful in propagation operations, where propagation from the image edge over the background should stop at object borders.

4.2.6 Connectivity of basic objects in an N dimensional image.

In the previous sub-section we justified that the best choice for background connectivity in any dimension is the lowest connectivity. In this subsection we will discuss the connectivity that complies best with a basic object $O^d_{N,\tilde{N}}$. For example, the best connectivity for curves in X_2 and X_3, for surfaces and volumes in X_3.
If we assume that the only objects in X_N are basic objects then the following model is used to describe an image X_N, its foreground objects and its background (see figure 4.2.11 for a 2-D case):

Proposition: *Each basic object $O^d_{N,\tilde{N}}$ is enclosed by a basic object $O^d_{N,\tilde{N}-1}$.*

It is assumed that all objects in an image are defined on a finite closed interval. In fact all basic objects in an image are segments, e.g. line segments, surface segments. Hence a line or curve segment is enclosed by end points, a surface segment is enclosed by a contour (a curve) and a volume segment is enclosed by a shell (a surface). In analogy with calculus, the terms open interval objects for objects without their boundaries and closed interval objects for objects including their boundaries are used. In concurrence with this the terms shell and core of an object will be used.

Proposition: *The boundary of a basic object $O^d_{N,\tilde{N}}$ can be envisioned to have another connectivity than the basic object itself.*

Figure 4.2.11 shows a basic plane in X_2 which can be divided into its boundary, a closed contour, and its core. Four situations are shown:

a) A 4-connected boundary with a 4-connected core.
b) An 8-connected boundary with a 4-connected core.
c) A 4-connected boundary with an 8-connected core.

d) An 8-connected boundary with an 8-connected core.

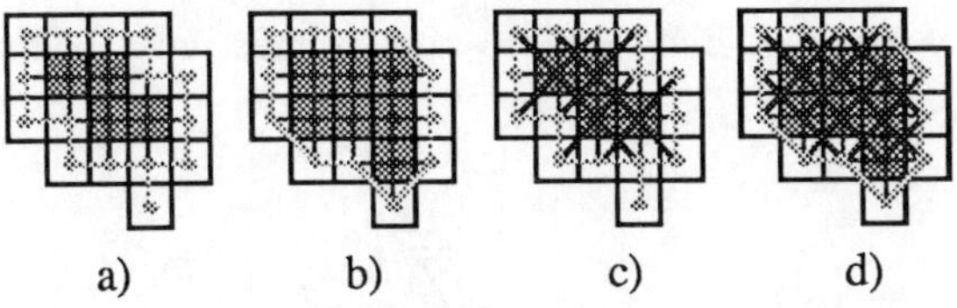

Figure 4.2.11 A basic plane in X_2. The boundary of the basic plane is a curve and may be envisioned to have another connectivity than the interior.

Note that a connectivity is a property that can be chosen or defined. Situations a) and c) show that due to this choice, the boundaries became compound objects (a small line segment sprouts at the bottom of the plane). Note, that in the upper left corner of situations b) and d) the boundary, although 8-connected, is locally 4-connected, a property that is allowed for 8-connected curves. A possibility would be to consider the three upper left pixels as a closed 8-connected curve adjacent to and connected with the basic plane. However, if we define closed curves as having at least one image element in its enclosed space, this viewpoint is invalid.

With the application in mind (thinning) a proper choice should be made for the connectivity of each basic object. This choice depends on the properties that we require for the objects.

Postulation: *The acquisition of a physical world image down to a binary image is such that all basic objects are closed structures.*

Note that this specifically excludes sponge-like structures for which it can be argued that they are volumes or planes with a low density. If these structures occur in an image they will be regarded as volumes, planes or surfaces with enclosed background objects (e.g. points, lines) or more commonly: Holes. Identically, dashed lines or curves are not recognized as one structure but merely as a number of unrelated small structures. Again, the boundaries of the objects at the side of the enclosed background objects are also assumed to have a lower intrinsic dimension.

We now make the assumption that the real properties of a basic object cannot be found at its boundary but only in its core. Then, if sponge-like structures are not allowed, it is not useful to choose a too high connectivity for the core of a basic object. See for an example figure 4.2.12.

Figure 4.2.12 Choosing a connectivity for the core of a basic object.

For the objective of keeping its image elements connected, a connectivity higher than 4 is not useful for the core of a planar object in X_2. Equally, a connectivity higher then 6 is not useful for the core of a volume in X_3 and a connectivity higher than 18 is not useful for the core of a surface in X_3.

An argument to maintain the highest useful connectivity is found in the curvature that basic objects are supposed to make. The higher the connectivity, the smaller the turning circles that can be made. This is depicted in figure 4.2.13. The next sub-section establishes a relation between a piecewise polynomial geometric description of a basic object and its connectivity. Being ahead of this, a conclusion of this sub-section is that the higher the connectivity, the higher the degree and the richness of the degree of the describing polynomials. Which is consistent with the curvature property.

Figure 4.2.13 Curvature potential per connectivity.

The previous discussion can be phrased with the following

Observation: *An appropriate choice for the connectivity of a basic object* $O^{d}_{N,\tilde{N}}$ *is the* $\{G_N^{\sqrt{N-\tilde{N}+1}} \mid (1 \leq \tilde{N} \leq N)\}$ *connectivity.*

Or: A justified appropriate choice for a plane in X_2 is the 4-connectivity and for a curve the 8-connectivity. A justified appropriate choice for a volume in X_3 is the 6-connectivity, for a surface the 18-connectivity and for a space curve the 26-connectivity.

A consistent system of foreground and background connectivities can be obtained by arguing that each foreground object is enclosed by a background object. As the most appropriate choice for a background connectivity was found to be the lowest connectivity, each 18-connected foreground surface in X_3 is assumed to be enclosed by a 6-connected background surface. Note that as the shell of a volume is assumed to be a surface too and hence 18-connected, closed interval volumes are also enclosed by a 6-connected background surface. In $X_{3,}$ curves and points alike are enclosed by a 6-connected background surface. In situations in the image where the objects are on the narrowest possible distance together, a 6-connected background surfaces is located in between those objects. Equivalently background passages through volumes are 6-connected in X_3.

There is no fundamental objection against assuming that the rest of the background stuffing in the image is background volume core and hence may have another connectivity, e.g. 26-connected. However, as operations on the image are supposed to be performed only on the foreground, this assumption has no practical implication. Note that if it is necessary to perform an operation on the background, the definitions of foreground and background can be reversed leaving the connectivity scheme intact.

For X_2 a suitable set of connectivities is: Planes 4-connected, curves 8-connected and background curves 4-connected. Background planes may be assumed 8-connected.

From the latter we learn that the inverse of this connectivity scheme also makes sense: Planes 8-connected, curves 4-connected, background curves 8-connected and background planes 4-connected. This scheme is a perfectly useful one, albeit that the object curvature that can be realized with this scheme is less in comparison with the inverse scheme.

Note that other schemes are also possible:

In section 4.1 a skeleton was proposed that was based on the erosion of 8-connected contours from 4-connected planes, but the final connectivity of the remaining skeleton (and hence of all non-border curves in the image) was 4-connected.

Concluding: Connectivities can be justifiably chosen resulting in a connectivity scheme. A set of tools can be chosen in accordance with this scheme. If the scheme is not chosen explicitly, the tools that are utilized on the image implicitly assume a specific connectivity scheme, which will show in the result of the operation.

4.2.7 The geometry of basic objects.

As an object primitive $P^{d}_{N,\tilde{N}}$ is the smallest possible entity still having the properties of the basic object, the core of a basic object can be formed by a tessellation of object primitives. For the non-curved object primitives $P^{d}_{N,N}$ only a single primitive is sufficient to tessellate a basic object $O^{d}_{N,N}$. Figures 4.2.3a and 4.2.3b depict structures that can be considered as the object primitives $P^{1}_{2,2}$, $P^{\sqrt{2}}_{2,2}$, $P^{1}_{3,3}$, $P^{\sqrt{2}}_{3,3}$, $P^{\sqrt{3}}_{3,3}$. All other basic objects $\{O^{d}_{N,\tilde{N}} \mid (1 \leq \tilde{N} \leq N-1)\}$ can be tessellated from various curved object primitives. Tessellation is assumed to be performed by starting with an initial object primitive $P^{d}_{N,\tilde{N}}$, followed by fitting another object primitive on each pixel of its neighbourhood. Note that the type of primitive that can be fitted depends on the occupancy of the other neighbourhood positions.
The questions that arise now are, how many object primitives $\{P^{d}_{N,\tilde{N}} \mid (1 \leq \tilde{N} \leq N-1)\}$ exist, and what is their shape?

In $\mathbb{R}_N$ normal geometrical relations can be used to describe basic objects. Basic objects with dimension $(0 < \tilde{N} \leq N-1)$ can be described using a set of $(N-\tilde{N})$ polynomials and for their boundary description one or more polynomial inequalities. For example with equation (4.2.3a) a 2D curve of degree M in $\mathbb{R}_2$ can be described. A set of 2 equations describes a point in $\mathbb{R}_2$. A set of inequalities analog to (4.2.3a) can be used to describe the surface area (partially) enclosed by the boundaries formed by these inequalities. With an equation of the form (4.2.3b) a surface in $\mathbb{R}_3$ can be described, with 2 equations of the form (4.2.3b) a space curve, with 3 equations a point in space and with a set of inequalities of the form (4.2.3b) the volume (partially) enclosed by the boundaries formed by these inequalities can be described.

$$x_2 = \sum_{m=0}^{M} a_m \cdot x_1^m \tag{4.2.3a}$$

$$x_3 = \sum_{m=0}^{M} \sum_{i=0}^{m} a_{m,i} \cdot x_1^i \cdot x_2^{m-i}$$

(4.2.3b)

As the basic objects were assumed to be tessellated from object primitives we make the following two definitions:

Definition: *An object primitive* $P^d_{N,\tilde{N}-1}$ *in a binary image* X_N *can be fully described with a polynomial. The polynomial is defined within the neighbourhood* M^3_N *which means that the function is defined within the interval* ($-1 \leq x_i < 1$) *with* $1 \leq i \leq N$.

Definition: *The core of a basic object* $\{O^d_{N,\tilde{N}} \mid (1 \leq \tilde{N} \leq N-1)\}$ *is described as a piecewise polynomial if it can be tessellated by object primitives* $\{P^d_{N,\tilde{N}} \mid (1 \leq \tilde{N} \leq N1)\}$.

From (4.2.3a) the polynomial for $P^{\sqrt{2}}_{2,1}$ can be extracted:

$$x_2 = a_0 + a_1x_1 + a_2x_1^2$$
$$\text{with } (a_1, a_2) \in \{\ 0,\ 0.5,\ 1 \mid a_1 + a_2 \in (0,1)\ \}$$

(4.2.4a)

which describes the curves $x_2 = 0$, $x_2 = x_1$, $x_2 = x_1^2$, $x_2 = 0.5x_1 + 0.5x_1^2$.

Note that a_0, the offset to the origin, is always 0 because all curves, surfaces, volumes, etc. are supposed to pass through the central element of neighbourhood M^n_N.

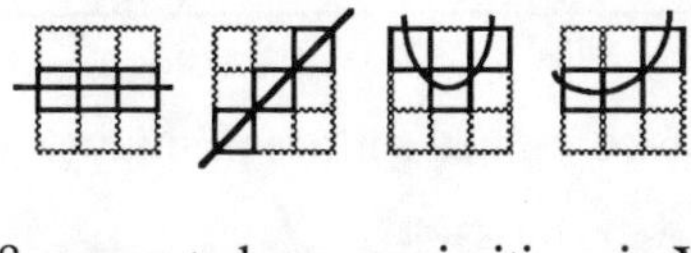

Figure 4.2.14 $P^{\sqrt{2}}_{2,1}$ or 8-connected curve primitives in X_2.

For $P^{\sqrt{3}}_{3,2}$ equation (4.2.3b) yields:

$$x_3 = a_{00} + a_{10}x_1 + a_{01}x_2 + a_{11}x_1x_2 + a_{20}x_1^2 + a_{02}x_2^2 + a_{21}x_1^2\,x_2 + a_{12}x_1x_2^2$$

with $(-1 \leq (a_{i,j}) \leq 1)$

and the resolution[3] of $(a_{i,j}) = 0.05$.

(4.2.4b)

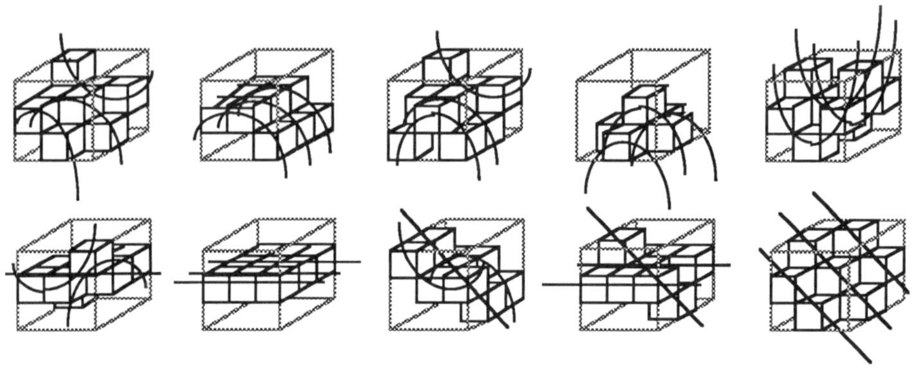

$x_3=$	$a_{10}x_1$	$+a_{01}x_2$	$+a_{11}x_1x_2$	$+a_{20}x_1^2$	$+a_{02}x_2^2$	$+a_{21}x_1^2x_2$	$+a_{12}x_1x_2^2$
$x_3=$	$0.5x_1$	$+0.5x_2$	$-0.25x_1x_2$	$-0.2x_1^2$	$-0.2x_2^2$	$+0.25x_1^2x_2$	$+0.25x_1x_2^2$
$x_3=$	$0.5x_1$			$-0.5x_1^2$			
$x_3=$	$0.5x_1$		$+0.25x_1x_2$	$-0.4x_1^2$	$+0.1x_2^2$	$+0.75x_1^2x_2$	$-0.25x_1x_2^2$
$x_3=$				$-x_1^2$	$-x_2^2$		
$x_3=$				x_1^2	$-x_2^2$		
$x_3=$						$-0.5x_1^2x_2$	$-0.5x_1x_2^2$
$x_3=$	0						
$x_3=$	x_1					$-0.5x_1^2x_2$	$-0.5x_1x_2^2$
$x_3=$	x_1						$-x_1x_2^2$
$x_3=$	x_1	$+x_2$					

Figure 4.2.15 $P^{\sqrt{3}}_{3,2}$ or 26-connected surface primitives in X_3.

[3] The resolution of the polynomial coefficients results from a matrix inversion operation on c_2 (see equation 4.2.6 in the sequel). The final form of an entry of $(c_2)^{-1}$ is a fractional term where the original entries of c_2 are located in a sum of products term in the denominator. While the range of the entries of c_2 is {-1, 0, 1}, the resolution of the polynomial coefficients depends on the number of column entries of polynomial coefficients.

The geometric description not only gives insight but is also needed in the sequel to geometrically intersect foreground and background object primitives. Hence, transformations between the geometric and the image element domains and vice versa are necessary. The image element description of an object primitive is given by a $\{x_1, x_2, x_3\}$ coordinate sequence.

The forward transformation is established as follows:

Using the matrix x_{12} of all $\{x_1, x_2\}$ coordinate combinations in M_3^3:

$$x_{12} = (-1, 1)\ (0, 1)\ (1, 1)\ (-1, 0)\ (0, 0)\ (1, 0)\ (-1,-1)\ (0,-1)\ (1,-1) \tag{4.2.5a}$$

and $a_{mi} = (a_{10}, a_{01}, a_{11}, a_{20}, a_{02}, a_{21}, a_{12})$, the vector of all polynomial coefficients of (4.2.3b) then this equation can be written as the matrix product:

$$x_3 = c_2 \bullet a_{mi} \tag{4.2.5b}$$

with:

$$c_2 = \begin{bmatrix} -1 & 1 & -1 & 1 & 1 & 1 & -1 \\ 0 & 1 & 0 & 0 & 1 & 0 & 0 \\ 1 & 1 & 1 & 1 & 1 & 1 & 1 \\ -1 & 0 & 0 & 1 & 0 & 0 & 0 \\ 0 & 0 & 0 & 0 & 0 & 0 & 0 \\ 1 & 0 & 0 & 1 & 0 & 0 & 0 \\ -1 & -1 & 1 & 1 & 1 & -1 & -1 \\ 0 & -1 & 0 & 0 & 1 & 0 & 0 \\ 1 & -1 & -1 & 1 & 1 & -1 & 1 \end{bmatrix}$$

For example $x_3 = a_{10} \cdot x_1 + a_{01} \cdot x_2 + a_{11} \cdot x_1 \cdot x_2 + a_{20} \cdot x_1^2 + a_{02} \cdot x_2^2 + a_{21} \cdot x_1^2 \cdot x_2 + a_{02} \cdot x_2^2 \cdot x_1$, yields for the point (-1, 1): $x_3 = -a_{10} + a_{01} - a_{11} + a_{20} + a_{02} + a_{21} - a_{02}$, the first row of c_2.

Equation (4.2.5b) can be depicted more spatially in the M_3^3 neighbourhood:

$-a_{10} + a_{01} - a_{11} + a_{20} + a_{02} + a_{21} - a_{12}$	$a_{01} + a_{02}$	$a_{10} + a_{01} + a_{11} + a_{20} + a_{02} + a_{21} + a_{12}$
$-a_{10} + a_{20}$	0	$a_{10} + a_{20}$
$-a_{10} - a_{01} + a_{11} + a_{20} + a_{02} - a_{21} - a_{12}$	$-a_{01} + a_{02}$	$a_{10} - a_{01} - x_1x_2 + a_{20} + a_{02} - a_{21} + a_{12}$

Table 4.2.2 Polynomial coefficients in the M_3^3 neighbourhood.

Equation (4.2.5b) gives the transformation from a polynomial description to description of an object primitive $P_{3,2}^{\sqrt{3}}$, a 26-connected surface primitive in an M_3^3 neighbourhood. It is clear that there is no unique mapping of a coefficient vector a_{mi} to an object primitive description. Only distinct coefficient vectors will pass exactly through the centers of the image elements. If all surfaces are not only allowed to pass through the center of an image element but also through the center $\pm$ 0.5, then many coefficient vectors and hence many surfaces may cross the same combination of elements yielding the same object primitive. Then x_3 must be rounded off to obtain a valid description ($x_3 \mid \{-1, 0, 1\}$) of the object primitive:

$$x_3^* = \text{sign}(x_3) \cdot \text{round}(\,|x_3|\,) \tag{4.2.5c}$$

The object primitive description is then given by $\{x_{12}, x_3^*\}$.
The inverse transformation, from the object primitive description to coefficient vector description a_{mi} is given by the matrix product:

$$a_{mi} = (c_2)^{-1} \bullet (x_3)^T \tag{4.2.6}$$

With $(c_2)^{-1}$ the matrix inverse of c_2 given by:

$(c_2)^{-1} =$

0	0	0	-0.5	0	0.5	0	0	0
0	0.5	0	0	0	0	0	-0.5	0
-0.25	0	0.25	0	0	0	0.25	0	-0.25
0.1	-0.2	0.1	0.3	0	0.3	0.1	-0.2	0.1
0.1	0.3	0.1	-0.2	0	-0.2	0.1	0.3	0.1
0.25	-0.5	0.25	0	0	0	-0.25	0.5	-0.25
-0.25	0	0.25	0.5	0	-0.5	-0.25	0	0.25

For 18-connected object primitives $P_{3,2}^{\sqrt{2}}$ the elements that are point-connected do not take part in its description (are "don't care"). The term $(+a_{21}x_1^2x_2 + a_{12}x_1x_2^2)$ in equation (4.2.3b) makes it possible to position the point-connected elements independently from other image elements at all possible point-connected positions. See also table 4.2.2. Consequently, for the 18-connected case these terms can be omitted and for $G_3^{\sqrt{2}}$-connected object primitives, equation (4.2.4b) yields:

$$x_3 = a_{10}x_1 + a_{01}x_2 + a_{11}x_1x_2 + a_{20}x_1^2 + a_{02}x_2^2$$
with $(-1 \leq (a_{i,j}) \leq 1)$
and the resolution of $(a_{i,j}) = 0.05$

(4.2.7a)

Likewise, for 6-connected object primitives $P^1_{3,2}$ the elements that are edge-connected do not take part in its description (are "don't care"). The term $(+ a_{11}x_1x_2)$ in equation (4.2.3b) makes it possible to position the edge connected elements independently from other image elements at all possible edge-connected positions. See also table 4.2.2. Consequently, for the 6-connected case this term can be omitted and for G^1_3-connected object primitives, equation (4.2.4b) yields:

$$x_3 = a_{10}x_1 + a_{01}x_2 + a_{20}x_1^2 + a_{02}x_2^2$$
with $(-1 \leq (a_{i,j}) \leq 1)$
and $(a_{i,j}) \in \{0, 1\}$

(4.2.7b)

Which yields $x_3 = x_1+x_2$, $x_3 = x_1+x_2^2$, $x_3 = x_1^2+x_2^2$.

Finally, for G^1_2-connected object primitives equation (4.2.4a) yields:

$$x_2 = a_1x_1 + a_2x_1^2$$
with $(a_1, a_2) \in \{ 0, 1 \mid a_1+a_2 \in (0,1) \}$

(4.2.7c)

or the curves $x_2 = x_1$, $x_2 = x_1^2$.

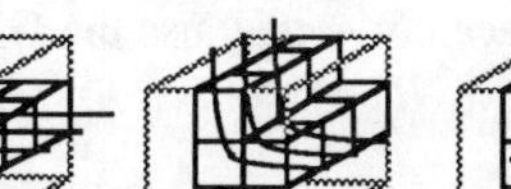

a) G^1_2 or 4-connected curve primitives in X_2

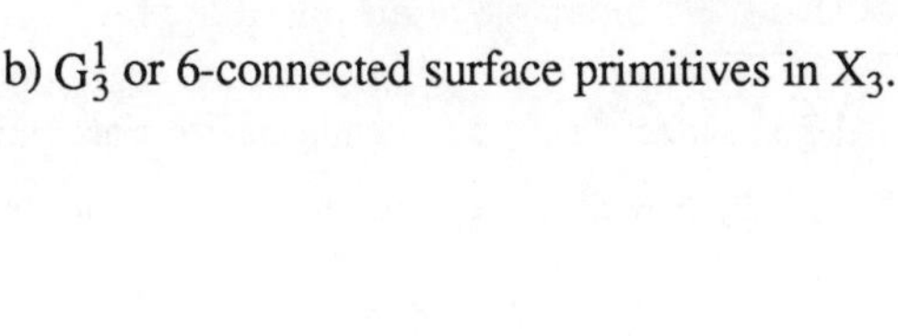

Figure 4.2.16 b) G^1_3 or 6-connected surface primitives in X_3.

However, due to the fact that a polynomial description cannot be given if more than one point is found for the dependent variable, the coordinate system must be rotated for the polynomial description of all G_N^1-connected object primitives.
This change of viewpoint does not alter the polynomial, apart from a scaling factor. By way of example, consider the last surface primitive of figure 4.2.16b: After rotating the coordinate system (x_3 axis from body diagonal to vertical) the points (1,0,0), (0,-1,0) and (0,0,1) rotate to ($\frac{1}{2}\sqrt{2}$, $-\frac{1}{6}\sqrt{6}$, $\frac{1}{3}\sqrt{3}$), ($-\frac{1}{2}\sqrt{2}$, $-\frac{1}{6}\sqrt{6}$, $\frac{1}{3}\sqrt{3}$) and (0, $\frac{1}{3}\sqrt{6}$, $\frac{1}{3}\sqrt{3}$) and the polynomial changes to $x_3 = \frac{1}{2}\sqrt{3}\,(x_1^2+x_2^2)$.

Summarizing, it can be observed that with the increase of the neighbourhood size the degree and the richness of the degree of the describing polynomial also increases.

- In a G_2^1 neighbourhood, some but not all, second degree curve primitives can be described.
- In a $G_2^{\sqrt{2}}$ neighbourhood all second degree curve primitives can be described.
- In a G_3^1 neighbourhood some, but not all, second degree surface primitives can be described.
- In a $G_3^{\sqrt{2}}$ neighbourhood all second degree surface primitives can be described.
- In a $G_3^{\sqrt{3}}$ neighbourhood some, but not all, third degree surface primitives can be described.

This leads to the conclusion that for the description of object primitives up to the second degree within a M_N^3 neighbourhood, the $G_2^{\sqrt{2}}$ neighbourhood is best suited for 2-D curves, and the $G_3^{\sqrt{2}}$ neighbourhood is best suited for 3-D surfaces. Note that if the image sampling density is so high that only linear patches can be expected locally (in the M_N^3 neighbourhood), one may always choose a lower connectivity.

The transformations for object primitives $\{P^d_{N,\tilde{N}} \mid (\tilde{N} < N-1)\}$ can be performed by using the transformations for $\tilde{N} = N-1$ and intersection. Note that if we intersect two $G_3^{\sqrt{2}}$ surfaces[4] the resulting space curve may use the $G_3^{\sqrt{3}}$ neighbourhood.

[4] Intersecting two second degree surfaces may lead to a third degree space curve (Roger and Adams 1976). (Observe a lawn tennis ball). However, M_3^3 is too small to show this phenomenon.

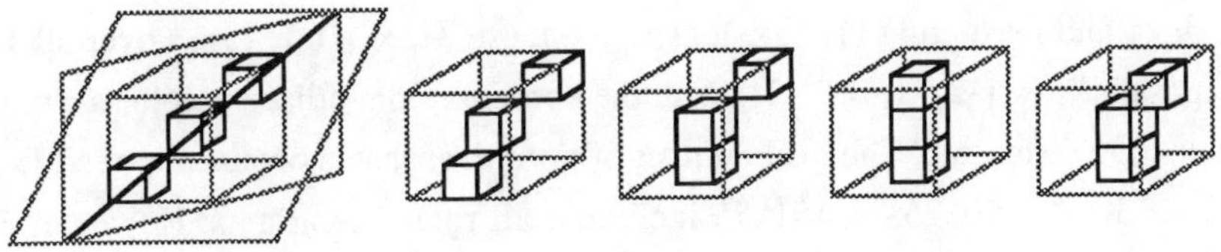

Figure 4.2.17 Examples of $G_3^{\sqrt{3}}$ or 26-connected curves in X_3.

4.2.8 Detection of basic objects in images.

In the previous subsection we presumed basic objects to be tessellated from object primitives and a relation was established between the geometrical descriptions and the image element descriptions of object primitives.

Presuming that all objects in an image are basic objects, and knowing all object primitives of all intrinsic dimensions, then, if we match each point in an image with all known object primitives, we are able to label each image element with its intrinsic dimension. Consequently, we are able to classify and detect each basic object in the image.

There are two ways to find a set of object primitives for a specific intrinsic dimension:

- By enumeration in the image element domain.
- By generation in the polynomial domain.

In both procedures, usually doubles are generated that have to be singled out afterwards. Moreover, the polynomial primitives are usually only generated for a single orientation. If the generated set of primitives is augmented with all its rotated (permutation of x_i) variants it can be used to detect any basic object in image X_N. If, however, only the unique topological different primitives of a set have to be described, all its rotated (permutation of x_i) and mirrored ($\pm a_i$) variants can be removed.

By way of example, the generation of all $G_3^{\sqrt{3}}$ surface primitives is described:

Choose an orientation (a rotation variant) by defining an axis in the M_3^3 neighbourhood, e.g. $\{x_1 = x_2 = 0\}$. This is effectuated by setting (0, 0, 0) to 1 and (0, 0, 1) and (0, 0, -1) to 0. The 26-connected surface can be found by permuting the columns (0, 1, x_3), (0, -1, x_3), (1, 0, x_3), (-1, 0, x_3) over all 1 out of 3 possibilities

{-1, 0, 1} and permuting (1, 1, x_3), (1, -1, x_3), (-1, -1, x_3), (-1, 1, x_3) over all 1 out of 5 possibilities {-2, -1, 0, 1, 2}, with the condition that adjacent columns remain face or edge connected. The total number of primitives that is obtained around {x_1=0, x_2=0} is $3^4 \cdot 5^4$ = 50625, or 151875 including all rotation variants. However, after removing the rotated and mirrored variants only 78 topologically different primitives can be distinguished.

The total number of 18-connected surface primitives obtained around {x_1=0, x_2=0} is 3^4 = 81, or 243 including all rotation variants. After removing the rotated and mirrored variants only 10 topologically different primitives can be distinguished. Note that in this case the columns (1, 1, x_3), (1, -1, x_3), (-1, -1, x_3), (-1, 1, x_3) are "don't cares".

In X_2, the number of topologically different 8-connected curve primitives is 4, and the number of topologically different 4-connected curve primitives is 2. See also Appendix A.

A final remark on tessellation and detection using object primitives:

After tessellating a basic volume with a 6-connected volume primitive, the resulting shell is an 18-connected surface. Conversely, when detecting a volume using a 6-connected volume primitive, only the core is detected and the remaining unlabeled voxels form the shell, an 18-connected surface. Similar relations can be found between all other connectivities, which links the connectivity choices as introduced in sub-section 4.2.6 even more.

4.2.9 Detection of compound objects in images.

The preceding subsection showed that basic objects $O^d_{N,\tilde{N}}$ can be detected if they are each matched with a set of object primitives $P^d_{N,\tilde{N}}$. So far, all non-foreground elements in $P^d_{N,\tilde{N}}$ have been considered background. However, in order to cover the description, tessellation and detection of compound objects, "don't cares" need to be used.

Definition: *A mask is an object primitive with element values* {1, 0, D}, *with* D *being "don't care". A set of masks will be referred to as* $S^d_{N,\tilde{N}}$.

With the introduction of "don't cares", forking and merging situations in compound objects can be detected. Forking involves forking of basic objects with identical intrinsic dimensions (e.g. a forking curve) and merging involves merging of basic objects with different intrinsic dimensions (e.g. a space curve emerging from a curved surface).

A fork in an object matches more than one mask, each mask matching one branch, the other branch being "don't care", and assuming the rest of the neighbourhood to be background. See figure 4.2.18 for a 2-D example. Situation a) depicts a piece of image X_2 with a forking curve. Masks b) and c) both match this situation, each on a different branch.

Figure 4.2.18 Detection of forking objects.

If we want to match all forking and merging situations a maximum number of "don't cares" needs to be introduced.

We assume that the connectivities for all intrinsic dimensions in foreground and background are chosen in accordance with the strategies of the previous subsections. In figure 4.2.18, e.g., we had chosen a $G_2^{\sqrt{2}}$ foreground curve surrounded by a suitable background, G_2^1 to prevent "leakage". Consequently the only background that matters is on 4-connected positions. Background on 8-connected positions is "don't care".

If we want to detect curves, the curve primitive should not transform into a plane primitive. This happens if the two line tiles of the curve primitive mutually connect and form a plane tile. For this reason, it should be prevented that the two non-center elements make contact. So the background should be at both sides of the curve and only one background element at each side is sufficient to prevent this contact.

Figure 4.2.19a depicts the two situations that meet these requirements.

By examining figure 4.2.19a it will be clear that if the central pixel changes from foreground to background, the 8-connected foreground curve is broken and a 4-

connected background curve is formed. For this reason these masks are called break point masks as they prevent topology breaking by the background.

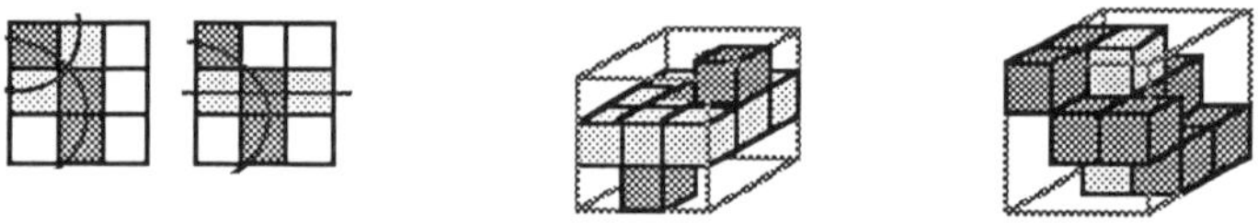

a) b) c)

Figure 4.2.19 Positions of the "don't cares".

Note that for space curves a 6-connected background ring (described as e.g. a background plane) should be applied to prevent contact between the two extreme voxels (cf. figure 4.2.19b). Likewise, for space surfaces only two voxels at each side of the surface are sufficient to prevent the surface primitive from becoming a volume primitive (figure 4.2.19c). Apparently the morphological task of the background is to preserve the foreground object primitive's dimensionality $\tilde{N}$.

> **Observation:** *Masks to detect basic objects* $\{O^d_{N,\tilde{N}} \mid (1 \leq \tilde{N} \leq N\text{-}1)\}$ *as well as the situations in which they fork, or merge with basic objects with other intrinsic dimensions, can be composed by unifying a foreground object primitive* $\{P^d_{N,\tilde{N}} \mid (1 \leq \tilde{N} \leq N\text{-}1)\}$*, with background object primitives while leaving the remaining image elements "don't care".*

The unification itself, determining all possible intersections of foreground and background object primitives, can be performed automatically by using the geometric equations of $P^d_{N,\tilde{N}}$ and their first and second derivatives. In X_3, e.g., using the normal to a curved surface and the tangent of a space curve, intersection of a foreground surface with a background curve is obtained by normal geometric procedures: The tangent of the curve should coincide (within a specific range) with the normal to the surface and the sign of the second partial derivatives should be equal. Note that the purpose of the coincidence is that the curve and surface really intersect and do not lay in the same plane. In this example, for the range a cone of 45° around the tangent can be used.

The result of this object primitive unification process is a set of break point masks $S^{d}_{N,\tilde{N}}$ that is suitable to detect basic objects in an image that may fork in its own intrinsic dimension or merge with other intrinsic dimensions.

4.2.10 Thinning as conditional erosion.

A local neighbourhood operation (LNO) on a source image X_N yielding a destination image Y_N can be defined as:

$$(\forall \vec{p} \in X_N)\ Y_{N\vec{p}} = f_{\vec{q} \in S}\ (X_{N\vec{p}+\vec{q}})$$

(4.2.8)

With S being the set of neighbourhood points connected to $\vec{p}$. S is called the structuring element for which in this case a square neighbourhood M^3_N is used. Note that if elements of S fall outside X_N, they belong to the edge ε_N (cf. subsection 4.2.2).

A recursive neighbourhood operation (RNO) on a source image X_N yielding a destination image Y_N can be defined as:

$$(\forall \vec{p} \in X_N)\ Y_{N\vec{p}} = f_{\vec{q} \in S}\ (X_{N\vec{p}+\vec{q}},\ Y_{N\vec{p}+\vec{q}})$$

(4.2.9)

Some of the possible techniques to update $(\forall \vec{p} \in X_N)$ image X_N are:

- Simultaneous or parallel. All points $\vec{p}$ are simultaneously updated. Normally only available on massive parallel machines.
- Serial or raster scan updating. Points are updated by raster scanning X_N from top-left to bottom-right. A variant of this approach is to scan X_N in all odd iterations from top-left to bottom-right and in all even iterations from bottom-right to top-left. Common in sequential algorithms and pipelined image processing machines.
- Queue and bucket updating. Candidate points $\vec{p}$ are stored in a queue and used in the next iteration (sometimes after -bucket- sorting on their distance to the border). The initial queue is mostly obtained by raster scanning the image. Queue updating reduces the unnecessary access of elements in an image. Only candidates that are likely to change are put onto the queue.

Thinning image X_N can be described in terms of an iterative and recursive neighbourhood operation on Y_N:

$$
\begin{aligned}
& Y_N^0 = X_N;\ i = 0; \\
& \text{repeat} \quad \{ \\
& \qquad (\forall \vec{p} \in Y_N)\ Y^i_{N\vec{p}} = f_{\vec{q} \in S}\,(Y^i_{N\vec{p}+\vec{q}}, Y^{i-1}_{N\vec{p}+\vec{q}}); \\
& \qquad i = i+1; \\
& \quad \} \\
& \text{until} \quad (Y^i_N = Y^{i-1}_N)
\end{aligned}
$$

(4.2.10)

The calculation sequence of $(\forall \vec{p} \in Y_N)$ is defined by the updating technique used in the calculation of the destination or skeleton image Y_N. On the one hand, the used updating technique influences the quality of the resulting skeletons in Y_N, but, on the other hand, contributions to the deviation due to the updating technique can be restricted to one image element for skeletons without end point conditions. (for explanation, see below).

Definition: *The structuring element for thinning a binary image* X_N *is represented by a set of mask-sets* S *with:*

$$S: \{ER_N, BP_N, EP_N\}$$

and:

ER_N: *The mask set specifying the erosion condition for intrinsic dimension* N.

BP_N : *The mask set specifying the break point conditions for intrinsic dimensions* $\tilde{N} = 0..N$.

EP_N : *The mask set specifying the end point conditions for intrinsic dimensions* $\tilde{N} = 0..N$.

(4.2.11)

Thinning is a conditional erosion: the erosion is performed if not only the erosion condition allows it, but also the break point and end point conditions. The break point conditions safeguard the topology and the end point conditions warrant the forming of the extremities of the skeleton.

Using this concept of mask-sets, the thinning process changes into a template matching problem: *If one of the masks of S fits, the deletion of the image element is* *__not__ permitted.*

Erosion is peeling a boundary from objects in X_N. The erosion itself can be performed using an erosion mask. For instance the masks of figures 4.2.3a and 4.2.3b (filled with foreground and “don’t cares” only) can be used to perform an 8- and 4-connected erosion in X_2 and a 26-, 18- and 6-connected erosion in X_3, respectively. In the previous sub-section it was however also shown that these masks are plane primitives and volume primitives and can be used to detect plane and volume core. In fact, by detecting the plane or volume core in a thinning process, they prevent the core from being deleted and leave only the shell for possible erosion. This is left for testing by the break point and end point mask-sets.

In the previous sub-section it was explained how a mask set $S^{d}_{N,\tilde{N}}$ for the break points of a single intrinsic dimension Ñ could be developed. For example mask set $S^{\sqrt{2}}_{3,2}$ as foreground can be merged with mask set $S^{1}_{3,1}$ as background to obtain a break voxel set $BP^{\sqrt{2}}_{3,2}$. (See figure 4.2.20 and Appendix A). The full break voxel set in X_3 is then $BP_N = \{BP^{\sqrt{2}}_{3,2}, BP^{\sqrt{3}}_{3,1}\}$.

Figure 4.2.20 Examples of masks from the break voxel set $BP^{\sqrt{2}}_{3,2}$.

Skeleton end-point conditions can be generated using a specific break point set and partially set its foreground elements to background. (In fact partial object primitives $Pp^{d}_{N,\tilde{N}}$ are used to detect the boundary situations of an object primitive.) As the skeleton end points are not supposed to fork nor to merge with other object primitives more “don't care” elements may be set to background. By doing this, skeleton artifacts are suppressed and other objects are kept on a specific distance.

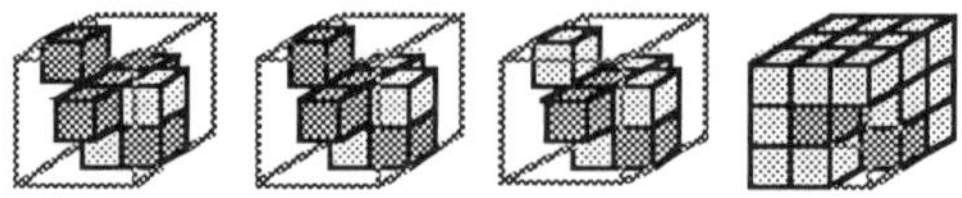

a) b) c) d)

Figure 4.2.21 Examples of masks from the end voxel set $EP^{\sqrt{2}}_{3,2}$.

Figure 4.2.21 shows some examples of an original 18-connected surface break point mask a) (i.e., preventing the erosion of an 18-connected full surface), the end point mask preventing the erosion of an 18-connected half-surface b) and the end point mask preventing the erosion of an 18-connected quarter-surface c). Mask d) is the same mask as c) but now with all possible "don't cares" replaced by background voxels. Note that two positions must remain "don't care" as either one can be foreground in the case of an 18-connected quarter surface, or both can be foreground in the case of a quarter surface rim emerging from a volume.

A final remark on skeleton end point conditions:
Generally, a better control over the sprouting of the skeleton ends can be obtained by either smoothing the surface of the objects before skeletonization or using larger neighbourhoods such as M^5_N. This gives more opportunity to control the situations in which skeleton ends should start to grow.

Figure 4.2.22 shows the masks ER^1_2, $BP^0_{2,0}$ and $EP^{\sqrt{2}}_2$ for the 2-D skeleton defined by $S = \{ER^1_2 \vee BP^{\sqrt{2}}_{2,1} \vee BP^0_{2,0} \vee EP^{\sqrt{2}}_2\}$.

erosion ER^1_2, single pixel $BP^0_{2,0}$.

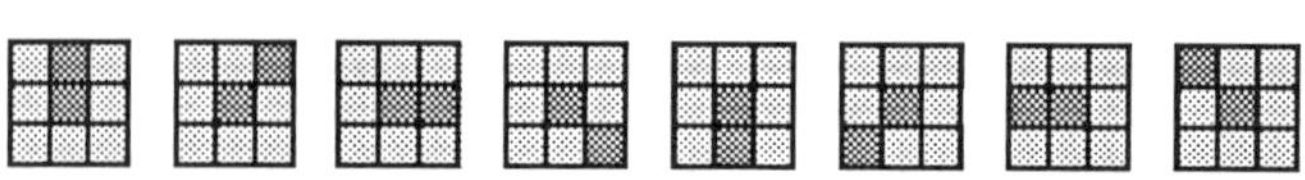

Figure 4.2.22 and end pixel mask set $EP^{\sqrt{2}}_2$.

4.2.11 Logic minimization and logic operations on mask sets.

For fast implementations, especially for high dimensional images, the $BP^{d}_{N,\tilde{N}}$ set and the $EP^{d}_{N,\tilde{N}}$ set should be as small as possible. In general the number of attempted matches is the product of the numbers of masks in foreground and background set. By applying logic minimization techniques, however, it appears that many background situations are already covered by other masks from the set.

For example merging $P^{\sqrt{2}}_{2,1}$ with $P^{1}_{2,1}$ results in a set $BP^{\sqrt{2}}_{2,1}$ in which each foreground mask only appears once. See figure 4.2.23. That this is not a general truth is shown in Appendix A, where the $BP^{\sqrt{2}}_{3,2}$ set shows that after minimization two masks ad_1 and ad_2 remained: One foreground with two background possibilities. Nevertheless, a substantial reduction can still be obtained using logic minimization. The number of masks of figure 4.2.23 was reduced by logic minimization from 38 possibilities to the 16 that are shown.

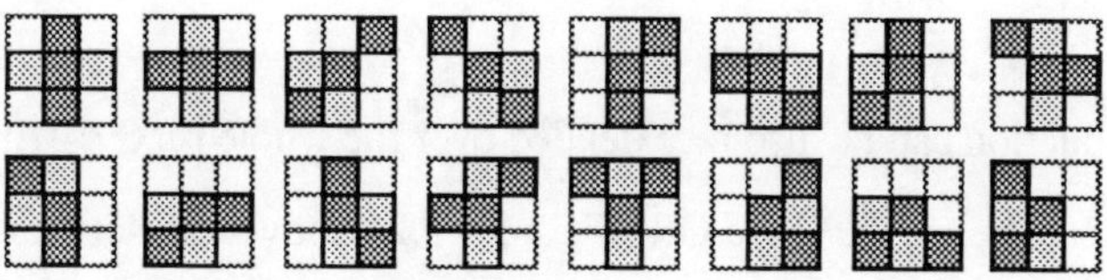

Figure 4.2.23 The break pixel set $BP^{\sqrt{2}}_{2,1}$.

Minimization is possible because mask sets can be expressed as logic equations. A mask from a set can be written as a logical AND term. The first mask of figure 4.2.23, e.g., can be described as: $(\overline{E} \wedge N \wedge \overline{W} \wedge S \wedge C)$, with $\wedge$ = AND, C= Center, E = East, $\overline{E}$ = NOT East, etc. Because "if one of the masks of $BP^{d}_{N,\tilde{N}}$ fits" can be translated in the logic OR of the members of the set, the whole set $BP^{d}_{N,\tilde{N}}$ can now be seen as a logic equation in the canonical form (AND-OR-form).

Consequently, minimal sets can be obtained by performing a logic reduction on the set using a logic reduction algorithm such as ESPRESSO (Brayton et al 1984a, 1984b). The minimal set resulting from such an algorithm is, however, not unique: other minimal sets may perform the same operation.

Applying other more common logic operations reveals some more properties of the sets such as:

Performing a logic inversion on the neighbourhood with the exception of the central pixel, the $BP^{\sqrt{2}}_{2,1}$ mask set, used for an 8-connected skeleton, can be transformed into a $BP^{1}_{2,1}$ set, for a 4-connected skeleton. Note, that the 1's (foreground) are replaced by 0's (background) and vice versa, the don't cares remain in place. Figure 4.2.24 shows the unreduced set $BP^{1}_{2,1}$ for only one orientation. For instance, the first four masks show a 4-connected horizontal foreground line segment intersected with respectively an 8-conected background, vertical line segment, diagonal line segment, obtuse curve and acute curve.

Figure 4.2.24 A part of the break pixel set $BP^{1}_{2,1}$ as obtained by inverting $BP^{\sqrt{2}}_{2,1}$.

The logic reduction can be also be extended over the whole set S = $\{ER^{d}_{N} \vee BP^{d}_{N,\tilde{N}} \vee EP^{d}_{N,\tilde{N}}\}$, with $\vee$ being the logic OR. Figure 4.2.25 shows the set that remains after reducing $S=\{ER^{1}_{2} \vee BP^{1}_{2,1}\}$: A thinning set that erodes 8-connected but yields a 4-connected skeleton.

Figure 4.2.25 The minimized set $S = \{ER^{1}_{2} \vee BP^{1}_{2,1}\}$.

Finally, if for instance the set $S=\{ER^{1}_{2} \vee BP^{1}_{2,1}\}$ is logically inverted, the masks of the set specify the elements of the image that should be deleted by each iteration. The latter is more in conformity with the classic way to describe thinning, but such a description lacks the possibility to visualize the (piecewise) polynomials of the resulting skeleton.

The mask set for the 3-D skeleton $S=\{ER^{1}_{3} \vee BP^{\sqrt{2}}_{3,2} \vee BP^{\sqrt{3}}_{3,1} \vee BP^{0}_{3,0} \vee EP^{\sqrt{2}}_{3,2} \vee EP^{\sqrt{3}}_{3,1}\}$ is given in Appendix A. Note that the masks are the unrotated and unmirrored versions.

The $BP^{\sqrt{2}}_{3,2}$ set and the $BP^{\sqrt{3}}_{3,1}$ set are reduced, which can be noticed from the fact that not all background possibilities occur in these sets. After mirroring, rotation and removal of double masks these sets can be used for thinning.
The 2-D skeletons are also summarized in Appendix A. Note that the mask set marked 2-D (a) is completely equivalent to the (Hilditch 1969) skeleton.
Appendix B finally shows the mask set that the 2-D (Arcelli et al 1975) skeleton appears to use in addition to the Hilditch skeleton set. This set was obtained by expanding both the Hilditch set and the Arcelli set to a table, extracting the differences, inverting this result, and minimizing it to a mask set again. The resulting mask set shows that this Arcelli skeleton does not generate a true 8-connected skeleton but matches besides for some 4-connected situations both in break pixels and in end pixels. As a result it is data dependent whether a 4-connected or an 8-connected fork is generated in the skeleton.

4.2.12 Thinning examples in 3-D.

To demonstrate the thinning process, some tube segments in X_3 have been taken (see figure 4.2.26a). The erosion condition ER^1_3 matches in each iteration with the volume core of the tube, leaving a $G^{\sqrt{2}}_2$ closed surface to be tested by the $BP^{\sqrt{2}}_{3,2}$ condition. The tube volume will be peeled away shell by shell until a one voxel thick $G^{\sqrt{2}}_2$ surface remains (figure 4.2.26b). The $BP^{\sqrt{2}}_{3,2}$ condition prevented the breaking of the surface connectivity and the $EP^{\sqrt{2}}_{3,2}$ condition prevented the surface from being eroded from its edges. Had the $EP^{\sqrt{2}}_{3,2}$ condition not been present, the tube segment would have been eroded to a $G^{\sqrt{3}}_3$ ring in the middle of the original segment. The vanishing of the ring was prevented by the $BP^{\sqrt{3}}_{3,1}$ condition (figure 4.2.26c).
The same process can be repeated for a cube with a hole. Original (figure 4.2.26d), eroded to closed surface (figure 4.2.26e, surface cut on 1/3 from the front for a better view) and after removal of $EP^{\sqrt{2}}_{3,2}$, erosion to a closed curve or wire frame (figure 4.2.26g).
Note that after erosion to a surface, the original erosion mask (that detected the volume core) has no role to play anymore when thinning further to a lower intrinsic dimension, such as from surface to wire frame. The surface break point masks take

over the role of erosion masks, as they are the ones that hit on the surface core and allow erosion to take place only on the boundaries of the surface.

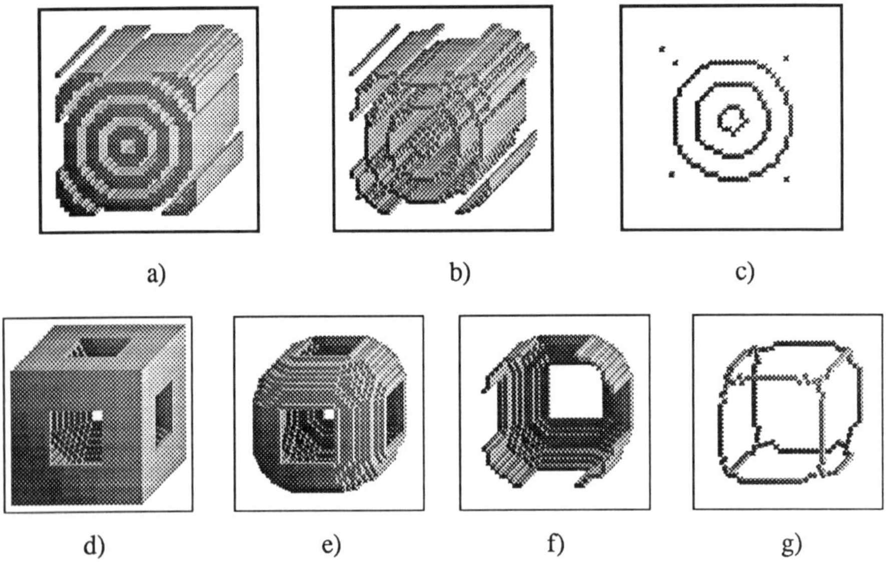

Figure 4.2.26 a) Original image (concentric pipes). b) 3D-skeleton with end-point condition. c) 3D-skeleton without end-point condition. d) Original image (cube with hole). e) 3D-skeleton with end-point condition. f) as e), but cut on 1/3 from the front for a better view. f) 3D-skeleton without end-point condition.

4.2.13 Parallel or sequential thinning and recursive neighbourhoods.

The masks shown in Appendix A can be implemented using a sequential update scheme or raster scan. A prerequisite for using the thinning set S in a sequential update method is that the set, with the exclusion of the erosion condition ER^d_N, should be applied in the recursive image Y^i_N. The erosion ER^d_N should be applied in image Y^{i-1}_N in order to keep the skeleton on the medial axis. The break point condition should be applied in image Y^i_N in order to be able to detect, using a M^3_N neighbourhood, new background emerging in the current thinning pass i, e.g., in the case of two pixel thick lines. In fact, detection of only the background in the image Y^i_N and only the foreground in the normal neighbourhood image Y^{i-1}_N is sufficient. We will refer to this as using the split neighbourhood.

Figure 4.2.27 shows the recursive neighbourhood M_2^3 to be used in a sequential update method in X_2 for a downward raster scan and an (elective) upward scan.

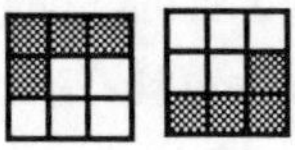

with: ■
= recursive image element
and: □
= normal image element

a) b)

Figure 4.2.27 Recursive neighbourhoods M_2^3 used for the downward raster scan, a) and (elective) upward raster scan, b).

The mask sets can be used in various implementations:

- For software implementation, the same recursive neighbourhood should be used. Fast processing can be obtained by skipping over the zeroes in the image and by ordering the masks by likelihood of appearance, as the mask testing OR-loop is broken on the first hit.
- A special hardware device (a Writable Logic Array) was developed to facilitate high speed pipelined thinning of 2D and 3D images in a few clock pulses per image element. See chapter 5.
- In massively data parallel architectures (e.g. meshes), usually one mask is used for all image elements simultaneously. In this case only the local neighbourhood can be used.

The difference between normal neighbourhood and recursive neighbourhood can be explained with an erosion. Applying the erosion mask in the normal neighbourhood shrinks the objects in the image in each iteration through the image. However, when using the recursive neighbourhood, the results in the same iteration are used and hence the erosion propagates over the objects until in one pass all object elements are eroded. Consequently, using the recursive neighbourhood for foreground elements of the skeleton mask sets too, will have a propagating effect. This may be useful in the skeleton 'without end pixel condition': Using the full recursive neighbourhood will result in a fast erosion of the skeleton ends, while using a split neighbourhood will result in the erosion of the skeleton ends only slowly (only one element in each iteration). The difference will be in position of the remaining skeleton, but for applications such as finding a path in a maze this is not important.

Figure 4.2.28 shows this effect on the example of the concentric tube segments. If the break point conditions $BP^{\sqrt{2}}_{3,2}$ are used in the full recursive neighbourhood (not only the background was tested in the recursive neighbourhood, but also the foreground), then the final skeleton lays on the far end of the image.
Note that in order to suppress the sprouting of too many skeleton branches, a better thinning performance is obtained if the end point mask sets are used in the normal neighbourhood only.

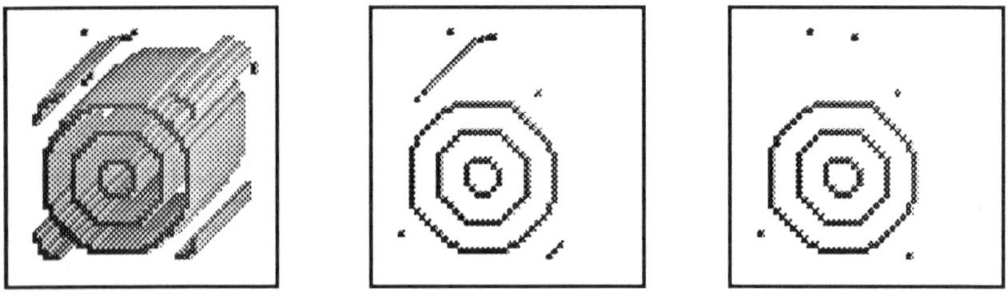

a) b) c)

Figure 4.2.28: Skeleton without end-point condition and the masks used in the full recursive neighbourhood, from the concentric pipes image of figure 4.2.26a.
After the first iteration the volumes are thinned to surfaces: a).
After the second iteration, the surfaces are thinned to lines: b).
After the third (final) iteration all non-closed lines are thinned to points: c). When used in the full recursive neighbourhood the resulting concentric pipes and rings are not situated in the middle of the original object but due to the recursive propagation over the volumes, surfaces and curves they are placed towards the ends of the original object . (Compare 4.2.28c with 4.2.26c.)

A **near Euclidean skeleton** can be obtained by applying a distance transform on the image, followed by thinning in the order of the found distances (Borgefors 1986, Verwer 1988, 1991).
In our approach, the (Verwer 1991) procedure can be followed, but instead of using the (Lobregt 1980) tables to prevent breaking of topology, the mask set $S=\{ER^1_3 \vee BP^{\sqrt{2}}_{3,2} \vee BP^{\sqrt{3}}_{3,1} \vee BP^0_{3,0} \vee EP^{\sqrt{2}}_{3,2} \vee EP^{\sqrt{3}}_{3,1}\}$ can be used. This approach has the advantage over the previously mentioned methods, that it allows thinning to lower intrinsic dimensions than $\tilde{N}$-1. For thinning without end point condition, the image

should be thinned until stable using $EP^{\sqrt{2}}_{3,2}$, followed by a constrained distance transform (CDT) on the resulting objects. Then a second thinning phase may be performed applying S without $EP^{\sqrt{2}}_{3,2}$, in the order of the then found distances, resulting in a wire frame-like object structure. When thinning is performed down to the lowest $\tilde{N}$, resulting in a point exactly in the center of the object (in terms of the used erosion metric), the CDT should be repeated and end point conditions should be removed from S for all intrinsic dimensions. Finally, the resulting skeleton for each object dimension $\tilde{N}$ lays on the medial axis.

4.2.14 Conclusions.

Driven by the aspiration to get insight in the principles of topology preserving thinning we have set up a general method for generating structuring element sets for thinning in N dimensions.

We have defined the connectivity between an image element $\vec{p}$ and an element $\vec{p}+\vec{q}$ as G^d_N-connected if $|\vec{q}| = d$ and $\max(q_i) = 1$ and showed its relation with the common indication of connectivity G: the number of elements in hypersphere S^d_N within neighbourhood M^n_N around image element $\vec{p}$.

We defined objects in images as compound structures composed of basic objects, whereas each basic object was defined as a non-forking group of image elements with an intrinsic dimension $\{\tilde{N} \mid (0 \leq \tilde{N} \leq N)\}$. We have defined object primitives as the smallest object still having the property of a basic object.

We discussed the connectivities of foreground and background objects, resulting in a scheme in which the best choice for the background connectivity is the lowest one, and the optimal connectivity for each intrinsic dimension $\tilde{N}$ is $G_N^{\sqrt{N-\tilde{N}+1}}$.

We have showed that mask sets could be established by intersecting proper foreground and background masks.

Thereafter, some thinning sets have been presented with some examples, and for sequential update techniques the role of the recursive neighbourhood was elucidated

We have showed that the mask sets could be minimized using logic minimization procedures and that common logic operations make it possible to manipulate thinning sets and to compare them. And finally a link with near Euclidean skeletons was indicated.

Straightforward software implementation of the 3-D mask sets takes about 3 seconds for a 32^3 image on a SUN-SPARC-2 (thinning the cube image to a wire frame). The masks were not ordered by likelihood of match, but skipping over zeroes was used.

Although the described method was developed to be dimension independent, it has not yet been implemented for thinning in a dimension higher than 3.

By using the method of this chapter, the original complexity of the 2^{27} (134,217,728) possibilities in a 3 x 3 x 3 neighbourhood has been brought back to a surprisingly low number of only 10 different, 18-connected surface primitives (yielding a minimized unrotated, unmirrored surface mask set of 20) and a number of only 13 different, 26-connected curve primitives (yielding a minimized unrotated, unmirrored curve mask set of 34). Consequently, with the erosion mask and single point mask, the skeleton without end-point conditions is described by only 56 topologically different masks. With mirroring and rotation the full skeleton with end pixel condition consists of 428 masks.

For 2-D, the original complexity of the 2^9 (512) possibilities in a 3 x 3 neighbourhood has been brought back to 4, 8-connected curve primitives, yielding a minimized, rotated and mirrored set of 26 masks for the skeleton with end point condition.

The enormous reduction of the original number of possibilities is largely due to the method of structuring the objects in the image. A further reduction was obtained by applying logic minimization.

5 Pipelined low level image processing.

In this chapter the design of an architecture for real-time low level image processing is discussed. This design served many goals: A basis for the comparison of massive data parallel SIMD architectures and instruction parallel pipelines. Experience in VLSI design as inevitable vehicle for future architectures. Research on mathematical morphology in multi-dimensional images, and finally, test object for VLSI design tools under development.

In section 5.1 the design of two special Logic Units for cellular logic processing is discussed. A Writable Logic Array for the parallel processing of sets of 3 x 3 binary structuring elements for the class of morphological operations, and a Tally circuit for the processing of the class of binary rank filters will be presented.

In section 5.2 the design aspects of real-time low-level image processing architectures are discussed from the designers point of view. Bottlenecks and design pitfalls on the road to a real-time system design are signalled.

In section 5.3 a design method for special computer architectures is presented. As control flow structuring may play an important role in the design of image processing architectures, the method focuses on asynchronous state machines to model this control.

In section 5.4 the architecture of a special VLSI circuit for Cellular Logic Processing is discussed. This Cellular Logic Processing Element (CLPE) is a device that is able both to operate as a single processor and as a Processing Element of a real-time binary image processing pipeline.

In section 5.5 the possibility is discussed to perform low-level greyvalue operations with CLPEs. A number of CLPEs can be stacked to form a Grey Value Slice and a number of Grey Value Slices can be cascaded to form a programmable pipeline for real-time low level image processing.

5.1 Devices for cellular logic processing.

5.1.1 Introduction.

In this section the design of two special Logic Units for cellular logic processing based on inexact matching will be discussed. A Writable Logic Array for the parallel processing of sets of 3 x 3 masks for the class of morphological operations and a Tally circuit for the processing of the class of binary rank filters is presented.

A special Cellular Logic Processing Element (CLPE) was designed (Jonker and Duin 1985, Kraaijveld et al. 1986) and realized (Jonker et al. 1988b) as a pipeline Processing Element, programmable for cellular logic operations based on sets of 3 x 3 structuring elements or masks. The CLPE is a full-custom VLSI circuit with as heart a Writable Logic Array (WLA) that performs the actual processing in the form of associative matches. This WLA is a newly designed and realized CMOS device for fast cellular logic processing and can be envisioned as a writable variant of a Programmable Logic Array (PLA).

In this section the WLA is described in detail as well as its generalization to a cascadable variant applicable in larger kernel cellular logic processing, three dimensional cellular logic processing and template matching.

5.1.2 The Writable Logic Array.

A Writable Logic Array was designed as a down loadable version of a Programmable Logic Array and a PLA can be considered as a generalization of a ROM. See figure 5.1.1a and 5.1.1b.

Let the matrix product between two matrices C <- A • B be defined as:

C <- A (+.*) B,

then the generalized matrix product can be defined as:

C <- A (f.g) B,

with f and g any dyadic arithmetic or logic function (Iverson 1980). Consequently, for C <- A • B, f is the addition function + and g is the multiplication function *.

As the addresses of a ROM are stored in the And_Plane, the address decoding of a ROM can be described as (see figure 5.1.1a) :

Tag <- And_Plane (∧.=) Address,

(5.1.1a)

with the logic equal function = being defined as $(A = B) \leftarrow (A \wedge B) \vee (\overline{A} \wedge \overline{B})$.

Note that the contents of the address space of the ROM is given by the matrix:

And_Plane <- Binary_Encode(0 .. $2^{addressbits}$ -1)

(5.1.1b)

The contents of the ROM are stored in the OR_Plane, the content retrieval can now be described as:

Content <- $(Or\_Plane)^{-1}$ (∨.∧) Tag

(5.1.2)

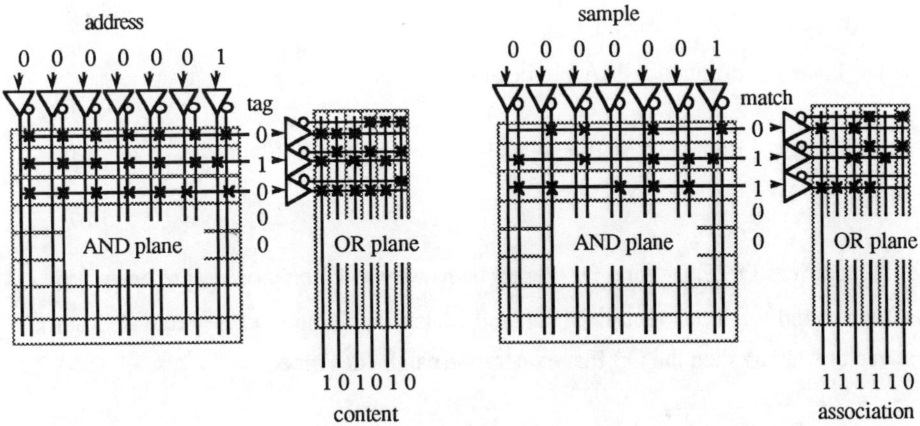

Figure 5.1.1 a) Read Only Memory b)Programmable Logic Array

Consequently, a ROM access can be described as:

Content <- $(Or\_Plane)^{-1}$ (∨.∧) (And_Plane (∧.=) Address)

(5.1.3)

In contrast with a ROM a PLA may contain don't cares; see figure 5.1.1b. The logic function (A = B) expands now to: $(A^t \wedge B) \vee (A^f \wedge \overline{B})$; A splits into two matrices A^t and A^f of which the elements $(A^t_{i,j}, A^f_{i,j})$ can be coded as:

$(A^t_{i,j}, A^f_{i,j})$ = { true: (1, 0), false: (0, 1), don't care: (0, 0), inhibit: (1, 1) }

(5.1.4)

The And_Plane operation of a PLA can now be described as:

$$\text{Match} \leftarrow [\forall\, i \mid \bigwedge_{j=0}^{\text{max\_col}} ((\text{And\_Plane}^{t}_{i,j} \wedge \text{Sample}_j) \vee (\text{And\_Plane}^{f}_{i,j} \wedge \overline{\text{Sample}_j}))] \quad (5.1.5)$$

Let the matrices And_Plane_t and And_Plane_f be column by column juxtaposed onto one matrix And_Plane* and the input vectors Sample and $\overline{\text{Sample}}$ be element wise juxtaposed onto one vector Sample*, then the And_Plane operation can be described with the logic matrix product:

$$\text{Match} \leftarrow \overline{(\text{And\_plane*}\ (\vee.\wedge)\ \overline{\text{Sample*}})} \quad (5.1.6)$$

The Or_Plane operation of a PLA can be described as:

$$\text{Association} \leftarrow [\forall\, j \mid \bigvee_{i=0}^{\text{max\_row}} ((\text{Or\_Plane}^{t}_{i,j} \wedge \text{Match}_j) \vee (\text{Or\_Plane}^{f}_{i,j} \wedge \overline{\text{Match}_j})) \quad (5.1.7)$$

Let the matrices Or_Plane_t and Or_Plane_f be row by row juxtaposed onto one matrix Or_Plane* and the input vectors Match and $\overline{\text{Match}}$ be element wise juxtaposed onto one vector Match* then the Or_Plane operation can be described as:

$$\text{Association} \leftarrow (\text{Or\_Plane*})^{-1}\ (\vee.\wedge)\ \text{Match*} \quad (5.1.8)$$

Consequently a PLA access can now be described as:

$$\text{Association} \leftarrow (\text{Or\_Plane*})^{-1}\ (\vee.\wedge)\ (\ \overline{\text{And\_Plane*}\ (\vee\cdot\wedge)\ \overline{\text{Sample*}}}\ )^{*} \quad (5.1.9)$$

Concluding, the inner product of the And_Plane and a row input vector yields a column vector of inexact matches. With this match vector, an association can be assembled using a matrix product of the Or_Plane and this match vector. Clearly a PLA can be seen as a Content Addressable PROM or an Associative PROM. Stressing these similarities further:

In a PROM[1] the and_plane is fixed and the or_plane is programmable.

---

[1] In a ROM the and_plane is fixed and the or_plane is fixed (factory programmed).

In a RAM the and_plane is fixed and the or_plane is writable.
In a PLA both the and_plane and the or_plane are programmable.
In a Content Addressable Memory, both the and_plane and the or_plane are writable.

For the CLPE only the inexact match is important and if any match occurs "the foreground" is associated. Consequently the WLA has a fixed OR (see figure 5.1.2) and the WLA read access function is described as:

$$\text{Result\_Out} \leftarrow \overline{(\text{And\_Plane*}\ (\vee.\wedge)\ \overline{\text{Nbh\_Opc*}})} \quad (5.1.10)$$

With Nbh_Opc* the concatenation of the input vectors: Neighbourhood vector, Recursive Neighbourhood vector, Mask plane bit (Image Z) and Operation code vector, totally juxtaposed with its inverse.

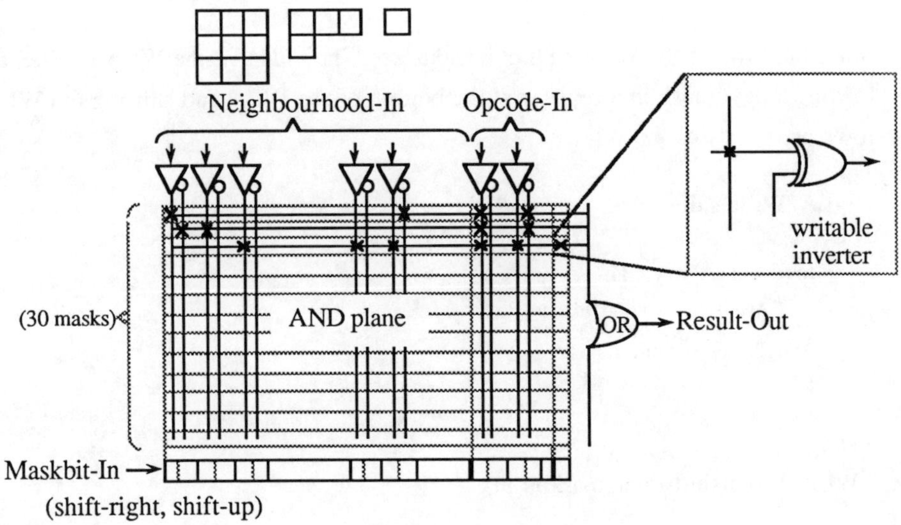

Figure 5.1.2 The Writable Logic Array.

As explained in section 4.1, masks are either SET or RESET masks and hence at the output of the And_Plane a column of writable inverters (EXOR gates) is placed. Consequently the WLA read operation becomes:

$$\text{Result\_Out} \leftarrow \bigvee_{\forall \text{rows}} \{\text{Invert} \neq \overline{(\text{And\_plane*}\ (\vee.\wedge)\ \overline{\text{Nbh\_Opc*}})}\ \} \quad (5.1.11)$$

or in shorthand notation:

Result-out <- WLA_Read (Nbh_Opc)

(5.1.12)

Equation (5.1.11) defines the device for the inexact neighbourhood match as defined in section 4.1 (equation 4.1.6).
Note that the column vector Invert is loaded together with the And_Plane. Note also that by loading an opcode bitpattern in addition with the mask data, masks can be grouped to a set. Masks with an identical operation code belong to the same set. By using don't cares in the operation codes, sharing of masks between sets is possible. For example, for skeletons with and without endpixel conditions, the breakpoint masks are shared which is realized by a don't care in their opcode (Jonker and Duin 1985). With the operation code provided at the WLA input, one operation can now be selected from the downloaded sets.

Since the rows of the WLA cannot be addressed individually, the WLA is filled by loading a new mask in a register at the bottom of the WLA and shifting the WLA rows up.

WLA_Write(shift-up):
{
for (i <- max_row .. 1)
{ And_Plane_i <- And_Plane_{i-1} }
And_Plane_0 <- Shift_Register
}

(5.1.13)

and:

WLA_Write(shift-right, maskbit-in):
{
for (i <- max_col .. 1)
{ $Shift\_Register_j$ <- $Shift\_Register_{j-1}$ }
$Shift\_Register_0$ <- maskbit-in
}

(5.1.14)

The mutual exclusion of the WLA_Read and WLA_Write operation was realized on the lowest level, allowing the WLA being loaded during process time. The CLPE uses a 2 phase clock scheme $\{\varnothing_1, \varnothing_2\}$. At each $\varnothing_1$ clock cycle a WLA_Write action (shift-right or shift-up) and at each $\varnothing_2$ clock cycle a WLA_Read action can be performed. This makes it possible that a new mask set is loaded during process time. The position of a mask does not influence its operation. However as masks are shifted out at the top, the user should keep a book of the positions of the masks in actual use.

The maximum row size of the WLA was a function of the WLA lay-out, the aimed clock-speed and the CMOS process[2] and was maximized to a size of 30 masks. This is large enough for most Cellular Logic Operations on 2 dimensional images. In order to limit the length of the input signal lines, the WLA had to be split up into two separate loadable sections of 15 masks each.

5.1.3 A General Purpose Writable Logic array.

Research on the subjects of cellular logic processing of 3 dimensional images, 4 x 4 kernel size cellular logic operations, and binary template matching as discussed in chapters 3 and 4 led to the question of the feasibility of large Writable Logic Arrays. For example for 3-D cellular logic operations within a 3 x 3 x 3 neighbourhood at least: $3^3 + (3^3 - 1)/2 + 1 + 6 = 48$ inputs (nnbh, rnbh, msk, opc) are required.
In template matching, where for instance a 16 x 16 neighbourhood around each pixel in a binary image is inexact matched with a template of ones, zeroes and don't cares, all templates can be matched in parallel using a Writable Logic Array. This feature matching requires 256 inputs by about 16 terms for a 16 x 16 neighbourhood.
However speed problems emanate when the WLAs are extended to a size larger than 15 terms by 20 inputs.

Figure 5.1.3 shows the logic diagram of a single WLA section of the CLPE. As can be seen, the AND plane of the WLA is implemented as a NOR plane. Using the inverted inputs and inverting the output, the required AND operation is obtained. As can also be seen in equation (5.1.6)
In the registerpairs in each row, the mask values are stored conform with the definitions of (5.1.4). In the last single register of a row the invert bit is stored, that is input to an exclusive-or gate. Each row is implemented using a pre-charged bus principle (Mukherjee 1986): On $Ø_1$ high the bus lines are pre-charged, the WLA inputs are sampled, the registers are clocked and the logic AND between registers and inputs is performed. On $Ø_1$ going low the pre-charging stops and the logic OR is

---

[2] Philips C5TH

performed; any input-register combination can pull the bus to ground. At $Ø_2$ high, the bus is assumed to be stable, the registers and the inputs are stable and the result can be clocked out. Note that this gives the opportunity on $Ø_2$ to shift the data of a row of registers one position up. Both the bottom row and the top row have a dynamic shift register to shift the WLA data serially in and out on $Ø_2$. The top register is used for test purposes. All row outputs are combined into the single logic

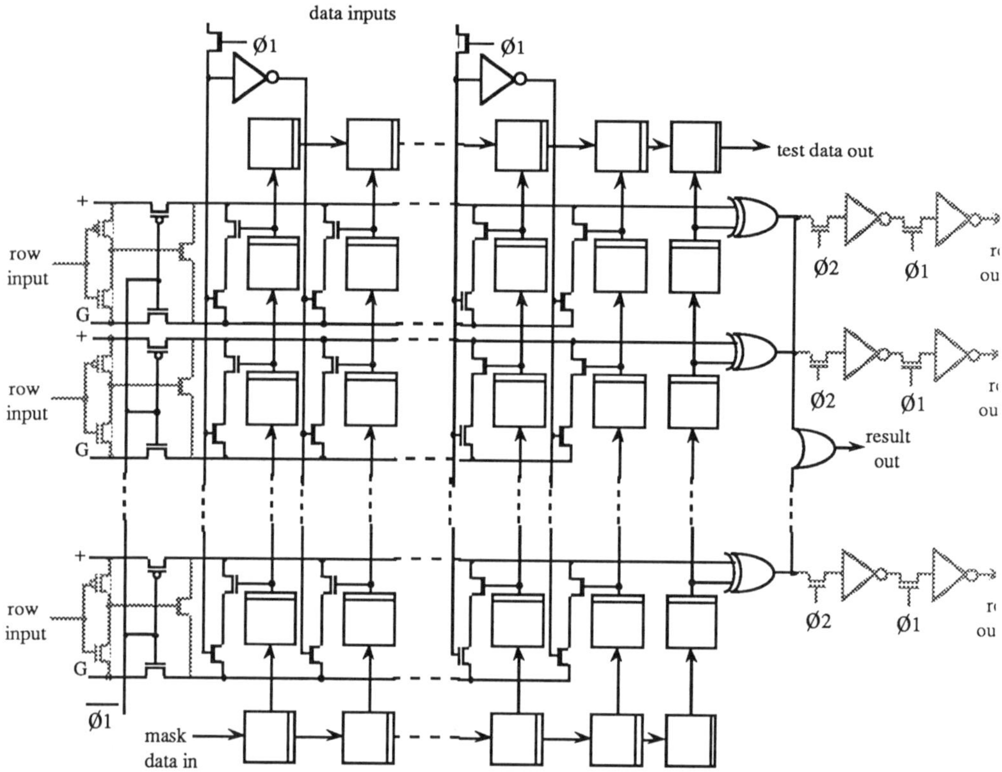

OR of the output.

Figure 5.1.3 The WLA logic diagram.

One WLA section of the CLPE of 41 columns by 15 rows of register cells has, with the current technology, more or less the maximum feasible size at a clock frequency of 10Mhz (Venema 1987, Hol 1987, Schot 1988). Note that 41 by 15 register cells is equal to a 20 inputs by 15 terms NAND plane.

The solution for larger WLA sizes must be found in cascading WLA sections. For this purpose a separate WLA chip was developed. The WLA chip was designed as a section of 17 by 8 register cells, 8 inputs by 8 terms, as a 40 pins device (Schmidt et al. 1988). Its maximum clock frequency was found to be 17 Mhz (De Zwart 1991). In figure 5.1.3 the shaded parts form the extensions to the original WLA that make the cascading of WLAs possible. The D-register at the row outputs allow the pipelining of WLAs, the input circuits at each row make it possible to pull the bus low if the row of previous WLA section had a low result.

Note that by operating two WLAs in parallel, more terms are obtained. Note also that as a WLA was realized as a NOR plane, an AND / OR WLA can be obtained by connecting the row outputs of one WLA to the data inputs of a second WLA, or:

$$\text{Output} \leftarrow \left\{ \text{Invert}_o \neq \left(\overline{(\text{Or\_plane}^{*-1})\ (\vee.\wedge) \left\{ \text{Invert}_a \neq \left(\overline{\text{And\_plane}^{*}\ (\vee.\wedge)\ \overline{\text{Input}^{*}}} \right) \right\}} \right) \right\} \quad (5.1.15)$$

Figure 5.1.4. shows how a 16 input, 16 terms, 8 output AND/OR WLA can be created by connecting six WLAs. The output values are available after 4 clock pulses. The input data must be stable during two clock pulses, hence halving the operation frequency. The OR can be used pipelined to the and plane, resulting in a delay of two clock pulses. Note that WLAs of 20 x 16 terms are feasible, the current WLA size of 8 x 8 was due to a 40 pins limitation[3].

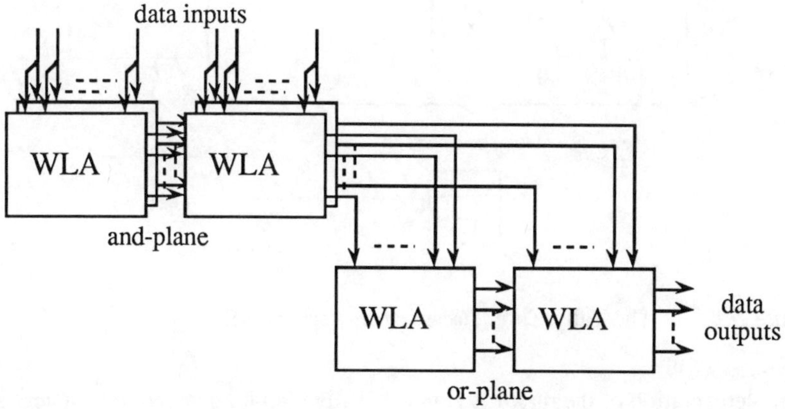

Figure 5.1.4 Cascading WLAs to obtain large and/or arrays.

[3] For prototyping; easy to produce and test.

5.1.4 The Tally circuit.

For the processing of the class of binary rank filters a WLA is not feasible as these type of filters is based on statistics and not on morphology. The number of masks to implement these filters is far too high (4.1.9) and a solution based on counting the number of ones in the 3 x 3 neighbourhood must be applied. To solve this problem a special solution based on counting was found using a Tally circuit (Mead and Conway 1980) that originally stems from relay-switching techniques (Caldwell 1959). See figure 5.1.5.

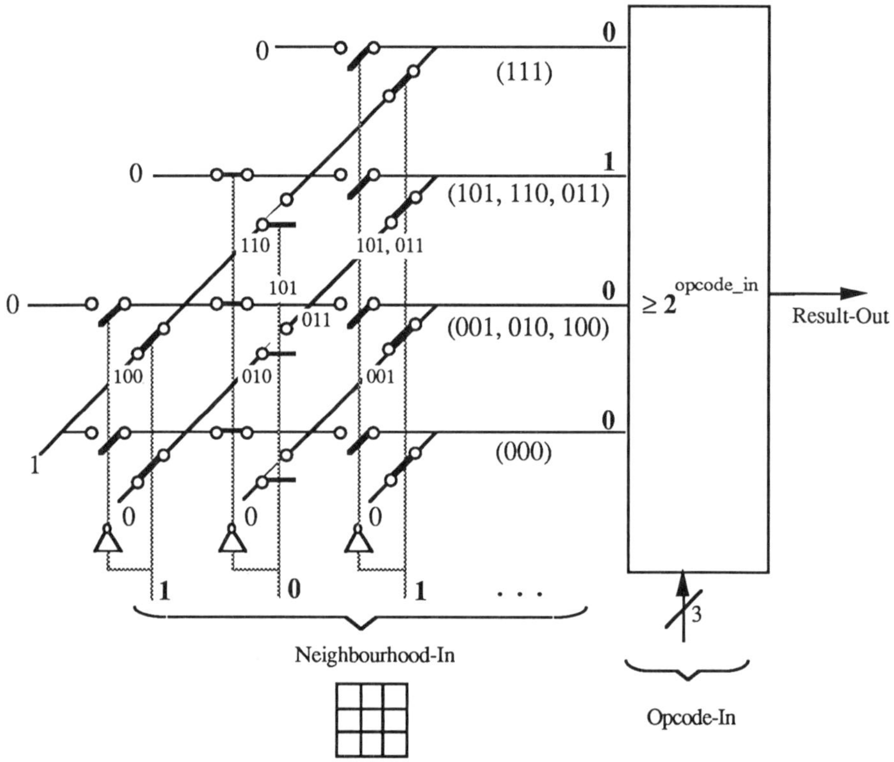

Figure 5.1.5 The Tally Circuit (adapted from Caldwell 1959)

The implementation of the rank filtering in a Tally circuit appeared to be a very small and clean solution for this counting operation due to the switch like behavior of MOS circuits. The area size of the Tally circuit on the chip was about 5% of the size of one WLA section.

5.1.5 Conclusions.

Cellular logic operations in 2 dimensional images can be defined as hit or miss transformations using sets of 3 x 3 binary structuring elements or masks that are used to perform an inexact match with every pixel of the image and its (recursive) neighbourhood. This match can be performed in parallel if the masks are stored as AND terms in a Writable Logic Array.

The Writable Logic Array is a variant of the Programmable Logic Array in the sense that it represents a down loadable AND plane cascaded to a row of downloadable inverters, terminated with a single OR gate. With the current CMOS technology a limitation was found at WLA sizes of 20 inputs by 15 terms at a frequency of 10 Mhz.

For several applications such as three dimensional image processing and template matching operations larger neighbourhoods are necessary. To overcome the current technology based size limitations, a cascadable variant of the WLA was made. With this device, extensions of terms is achieved without penalty in the propagation delay. The propagation delay is one clock cycle per added WLA for input extension. Moreover, AND/OR arrays can be created with the WLA. As the And-Plane and Or-Plane are pipelined, such a configuration does not lower the operation frequency.

For operations based on counting in a 3 x 3 neighbourhood, a very space efficient solution was found in a modified version of a Tally circuit.

5.2 Design aspects of real-time low-level image processors.

5.2.1 Introduction.

In many image processing applications, low-level image processing plays a pre-processing role for pattern recognition, scene analysis or measurements.

In product flow oriented industrial inspection tasks and in robot vision, the time constraints are very tight. By way of example, in a robot assembly cell, an average of six objects arrive simultaneously on a pallet at an input position in a robot workspace to be used for assembly by two robots (Jonker 1988c). The robots pick-up the objects from the pallet and assemble them on a workbench. The mean assembly duration is estimated to be 3 seconds which means that the robots are 3 seconds out of the field of view of the cameras at the input position and that for each object less than 3 seconds remains to calculate its 3 Dimensional Position and Orientation (3DPO) with an accuracy better than 0.5 mm. This situation becomes even worse if first the identity of all objects and their stable pose must be calculated within the first 3 seconds after the arrival of the pallet at the input position.

The maximum speed that can be obtained in low-level image processing using standard cameras is the video frequency. This is a technical barrier and hence under pressure of the applications as depicted above, at the same time a desirable goal to achieve. Moreover, many of the image processing systems will be placed in environments where no image processing experts are available and maintenance and calibration of certain (intermediate!) image processing steps still remains necessary. Then the possibility for real-time human-system interaction using a standard video interface does pay off.

This section starts with the set-up of a calculation model for a low-level image processor as a framework for discussion. Using the model, real-time behavior and the pipelining between image input, processing and image output is discussed. This is followed by a discussion on the characteristics of the four archetypes of low-level image processing architectures, the Square Processor Array (SPA), the Pyramid (PYR), the Linear Processor Array (LPA) and the Pipeline (PL). This discussion will **not** be done from a benchmarking or comparison point of view, but from a designer point of view. For a comparison between the architectural groups see (Komen 1990) and for the benchmarking of low-level image processing architectures, see (Preston

1989). The ability of the four architectural groups for real-time processing will be investigated and most bottlenecks and design pitfalls will be reviewed.

5.2.2 A theoretical processor for low-level operations.

Many object operations can be expressed as Recursive Neighbourhood Operations (Komen 1990). When excluding the class of object operations that can not be written as RNOs, the following observation on data structures for low-level image processing can be made:

Let the basic image datastructure be a two dimensional matrix of size N, the basic neighbourhood data structure be a two dimensional matrix of size n and the basic data structure for statistical data be a one dimensional vector or scalar. We will assume that this vector can be stored in a row or column of the image data structure and that the scalar can be stored in any image element. Let the basic data type for all aforementioned data structures be a data word of b bits.

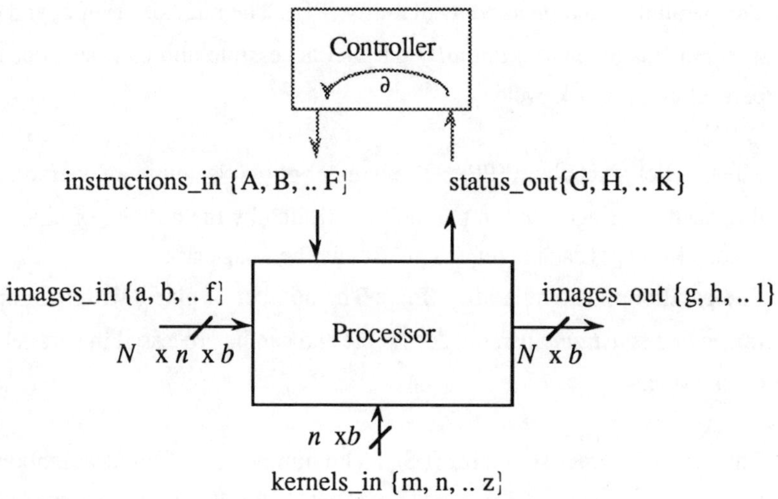

Figure 5.2.1 The basic processor for low-level image processing.

Let us assume that a theoretical processor as shown in figure 5.2.1 is capable to autonomously perform all necessary addressing to obtain instructions and to read and write data, and is able to process all low-level image processing operations as mentioned in section 2.2.1. The theoretical processor is assumed to be constructed from a number of *P* parallel and/or pipelined PEs. Let us assume also that within one idealized clock cycle the instructions {A, B, ..F} let the processor perform a transformation from input images {a, b, ..f} while using the neighbourhood kernels {m, n, ..z} to output images {g, h, ..l} and yielding the status information {G, H, .. K}. Parallel execution of instructions {A, B, ..F} can be achieved if some of the PEs of the processor are able to execute a different instruction. It is assumed that the global controller needs ∂ clockcycles to read the status {G, H, .. K} and produce a new set of instructions {A, B, ..F}.

The following (parallel resource claim - parallel resource availability) ratio, expressed in clock cycles, can now be distinguished:

1a. The parallel image-operand claim (PI_c): The number of image operands that need to be accessed in parallel, multiplied by the number of clock cycles needed to access each operand plus the number of clock cycles to perform the processing.

1b. The parallel image-operand availability (PI_a): The number of input and output operand images that is *actually* parallel accessible and can be processed in parallel in one clock cycle.

2a. The parallel pixel claim (PP_c): The number of pixels within an operand image, that need to be accessed in parallel, multiplied by the number of clock cycles needed to access each pixel. (PP_c = usually the image size).

2b. The parallel pixel availability (PP_a): The number of pixels within an operand image that is *actually* parallel accessible and can be processed in parallel in one clock cycle.

3a. The parallel kernel-set claim (PS_c): The number of different neighbourhood kernels that need to be accessed in parallel, multiplied by the number of clock cycles needed to access each neighbourhood kernel. (PS_c = usually the size of the kernel set).

3b. The parallel kernel-set availability (PS_a): The number of different neighbourhood kernels that is *actually* parallel accessible and can be processed in parallel in one clock cycle.

4a. The parallel kernel-connectivity claim (PK_c): The number of pixels within a kernel that need to be accessed in parallel to form the normal- and the recursive neighbourhood kernel (if applicable), multiplied by the number of clock cycles needed to access each kernel pixel. (PK_c = usually the kernel size).

4b. The parallel kernel-connectivity availability (PK_a): The number of pixels of a normal and recursive neighbourhood kernel that are *actually* parallel accessible and can be processed in parallel in one clock cycle.

5a. The parallel neighbourhood-connectivity claim (PN_c): The number of pixels within the input operand image(s) that need to be accessed in parallel to form a normal- and a recursive neighbourhood (if applicable), multiplied by the number of clock cycles needed to access each neighbourhood pixel in the input image(s). (PN_c = usually the neighbourhood size).

5b. The parallel neighbourhood-connectivity availability (PN_a): The number of pixels of a normal- and a recursive neighbourhood that are *actually* parallel accessible and can be processed in parallel in one clock cycle.

6a. The parallel bit claim (PB_c): The number of bits per pixel that need to be accessed in parallel, multiplied by the number of clock cycles needed to access each bit. (PB_c = usually the pixel word size).

6b. The parallel bit availability (PB_a): The number of bits per pixel that is *actually* parallel accessible and can be processed in parallel in one clock cycle. (The processor's word size).

7a. The parallel operations claim (PO_c): The number of operations that need to be performed, i.e. instructions that need to be executed, in parallel in one clock cycle, multiplied by the number of clock cycles needed to access each instruction (∂, the delay of the controller).

7b. The parallel instruction availability (PO_a): The number of instructions that is *actually* parallel available and can be executed in parallel in one clock cycle.

Note that:

1) For most operations a dyadic variant exists in which every pixel Y_p in the destination image is not only a function of the pixel X_p, (and its structuring element S and / or vectors p and q), but also from the pixel Z_p in a second source image Z. Even triadic operations exist (e.g. addition), in cases where the word

size of the processor is smaller than the word size of the image element and a carry-over bitplane is used.

2) In cellular logic processing, structuring elements can be implemented using a square neighbourhood or kernel and marking the pixels of the neighbourhood don't care on positions not belonging to the structuring element. This splits off a don't care kernel from the original kernel, yielding a transformation that uses two neighbourhood kernels preferably in parallel.
Moreover, situations occur in which a number of structuring elements can be applied in parallel on a single input image X to yield a single output image Y. For example eight anisotropical structuring elements can be used in parallel to obtain one isotropical transformation from X to Y.

5.2.3 Real time image processing.

If the parallelism claim is larger than the parallelism availability, sequential solutions must be found and provisions should be made where necessary to allow these solutions. For example:
For a larger image size than the parallel pixel availability, edge memory must be installed. For a smaller processor word size than the pixel word size, a carry bitplane must be installed. In the other cases no special hardware but intermediate storage is sufficient to handle sequential solutions.

Exchanging a parallel for a sequential solution gives for each of the 7 situations a delay of Ceil ($\beta_X \cdot PX_c / PX_a$) clock cycles, ß being an overhead factor, usually 1. The total delay of the sequential solutions, introduced to solve a lack of sufficient parallelism on any point in the processor, is given by:

$$D_{lsp} = \prod_{X = \{I,P,S,K,N,B,O\}} (\,\mathrm{Ceil}\,(\,\beta_{PX} \cdot PX_c / PX_a)) \qquad (5.2.1)$$

Let us assume that a low-level image processing task is built up from low level-operations and that low level operations are processed by executing a number of processor instructions. Then for a low-level image processing task a sequence of parallel instructions need to be performed by sending a stream of (parallel) instructions to the processor. Data dependent operation can be obtained by reacting on the status information that is returned by the processor after execution of each set of parallel instructions. Note that the fact that this stream with length l is sequential is forced by the algorithms of the task. It is evident that the stream may contain repetitions of the same instruction due to the fact that the controller controls the program control flow and hence the looping. Let us assume that the program load time is given by D_{plt}. Note that the delay between two instructions ∂, due to the time the controller needs to react on the status of the processor and to generate a new instruction, was modelled with I_a. The total delay for the operation becomes now:

$$D_{top} = l \cdot D_{lsp} + D_{plt} \tag{5.2.2}$$

In real-time image processing, a continuous stream of input images is sent to the processor. Let us assume that it takes a delay of D_{s2i} clock cycles to transform sensor data to image data and a delay of D_{i2d} clock cycles to transform image data to display data. Then the total delay for processing a single image is:

$$D_{im} = l \cdot D_{lsp} + D_{plt} + D_{s2i} + D_{i2d} \tag{5.2.3}$$

Note that with D_{i2d} also uploading to host memory can be modelled.
If the continuous stream of input images has a frequency of f images per second or $t = 1/f$ seconds of processing time available for each image, then for real-time low level image processing:

$$D_{im} \leq t \tag{5.2.4}$$

To narrow down the concept of real-time image processing and in order to make yet some form of comparison possible, the assumption is made that a frame grabbing unit grabs images from the video signal and places a 512^2 image of square pixels in a framebuffer (VI). As a normal interlaced video signal is cumbersome to operate on, it is assumed that a frame buffer is needed anyhow to make the signal non-interlaced.
The assumption is made that D_{im} is larger than the frame sync time, so the time between two frame sync signals (40 msec) is typically the time in which the low-level image processing task should be performed. The standard pixel frequency and PE frequency is, where necessary, assumed to be fixed to a convenient but realistic value

of 10 Mhz. The assumption is also made that a video output buffer (VO) exists from which the data is displayed or transported to a host following each frame sync. Figure 5.2.2a shows that real time processing is possible if $D_{im} \leq t$. Figure 5.2.2b shows that the operation frequency halves as soon as $D_{im} > t$.

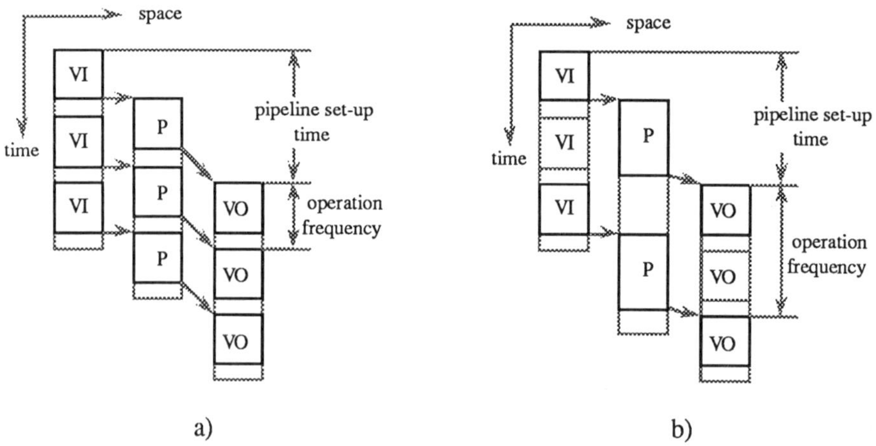

Figure 5.2.2 a) Real time processing with an SPA
b) Processing on half the real time frequency

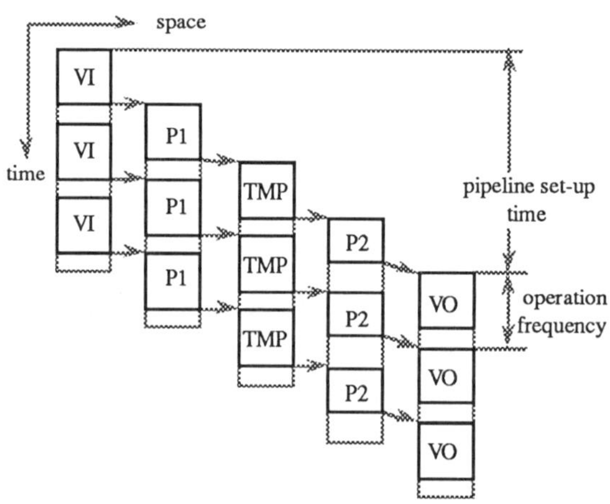

Figure 5.2.2c Pipelined Processing using a second set of PEs.

Figure 5.2.2c shows that a second set of PEs, memory between the sets and division of the task in two consecutive parts, makes operation pipelining possible. This causes an initial start-up delay but does not lower the operation frequency.
Note that the image_I/O-processing pipelining as shown in figure 5.2.2, although evident, must be adequately secured in the system design. For example in the (Stonefield 1986) version of the CLIP4 SPA, the Video Input, Processing and Video Output are controlled by separate UNIX kernel calls, with as a consequence that in worst case the time of many video frames is lost for both video input and video output, due to the non real-time nature of UNIX.
If the pipelining of data-I/O and processing is properly performed then $D_{im} = D_{top}$.

In the next subsection we will discuss some archetypical low level image processors and their ability to perform real-time processing. The various solutions found in spatially and temporal parallelism are elucidated, using (5.2.1) to (5.2.4) to signal possible bottlenecks and pitfalls. The low-level architectures are grouped in:

1) Square Processor Arrays (SPAs).
2) Pyramids (PYRs).
3) Linear Processor Arrays (LPAs).
4) Pipelines (PLs).

The discussion will be held using two algorithms:

Anchor skeletonization:
This algorithm is a dyadic iterative cellular logic operation. The number of operations is typically half the image size *N*. Objects may have as a maximum size the image size itself. The algorithm is performed using 8 masks with ones, zeroes and don't cares. If a recursive neighbourhood is used, the 8 masks can be matched in parallel, if only a non-recursive neighbourhood is used the 8 masks should be matched sequentially.

Shading correction in a 21 x 21 neighbourhood.
This algorithm is a monadic greyvalue operation performed as the original image minus (10 iterations of a 4 connected Local Maximum transformation of the original followed by 10 iterations of a 4 connected Local Minimum transformation).

5.2.4 Square Processor Arrays.

In massive data parallel machines the data is processed in a spatially parallel way by a massive number of parallel operating PEs, see figure 5.2.3.

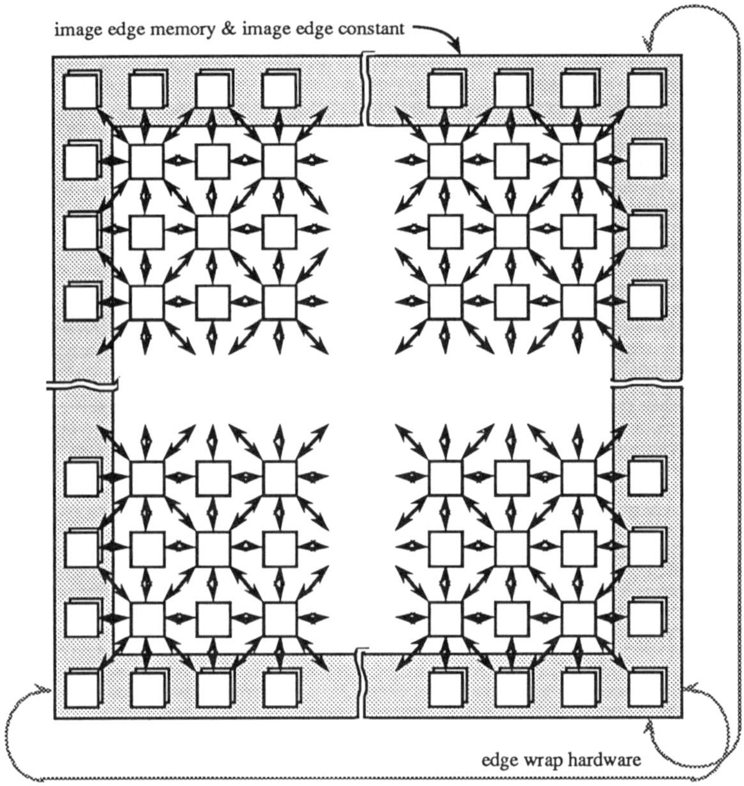

Figure 5.2.3 The data structure and PE interconnection structure of a SPA.

Clearly at Square Processor Arrays the concept of the system designer is to make $\mathrm{Ceil}(\beta_{PP} \cdot PP_c / PP_a) = 1$. In SIMD type machines the instructions for a certain operation are provided sequentially to all PEs by a global controller. As one of the most obvious data structures in image processing is the two dimensional matrix of data elements that constitute the image, the two dimensional grid of PEs with each PE connected to its 8 nearest neighbours, allowing fast retrieval of 3 x 3 filter kernels, is the most adequate data interconnection structure as figure 5.2.3 shows. Note that this structure implies both a hard wired 3 x 3 filter kernel structure as well as an *N* x *N* image data matrix structure. The edge of the image is ideally both presettable to a constant value and loadable from edge memory to allow scanning with a smaller array over a larger image. If larger images than the array size have to be processed, edge wrap hardware is almost indispensable. The memory organization of a SPA is organized in a pile of memory planes, whereas each word of the plane is only accessible by the underlaying processor (see figure 5.2.4).

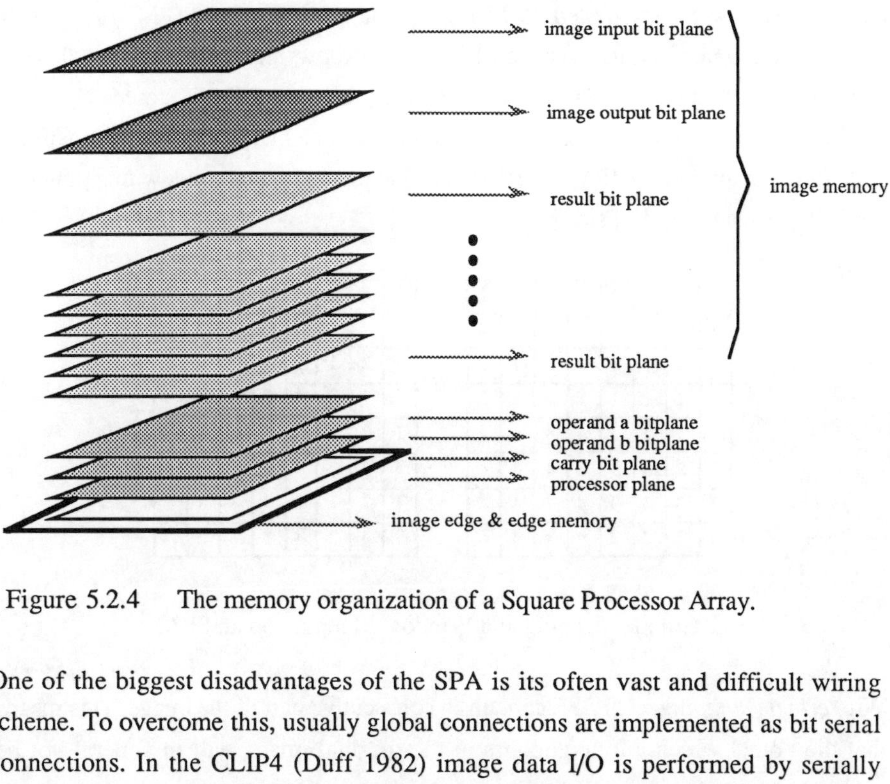

Figure 5.2.4 The memory organization of a Square Processor Array.

One of the biggest disadvantages of the SPA is its often vast and difficult wiring scheme. To overcome this, usually global connections are implemented as bit serial connections. In the CLIP4 (Duff 1982) image data I/O is performed by serially shifting data in and out through the A plane. Hence for both D_{s2i} and D_{i2d} the number of involved clock cycles is *P*, the number of PEs, divided by 32 when a 32 bits bus is used to load the A plane. Or: $D_{s2i} = D_{i2d} = 16 \times 512 = 8192$ clock cycles for a 512^2 image. D_{plt} is assumed to be zero as instructions are loaded one by one. Note that most recent SPAs have provisions in the form of a separate data I/O register to enable pipelining of data I/O and processing (see also chapter 3).

The first component of D_{lsp} is: Ceil ($\beta_{PI} \cdot PI_c / PI_a$), the delay ratio due to sequential image operand access. Assume a system with 3 input operands; an A-plane, a B-plane and a Carry (bit)plane and 32 output planes (D). If A, B, C and D can all be

accessed in parallel then Ceil ($\beta_{PI} \cdot PI_c / PI_a$) = 1. This is usually not the case. For the skeletonization algorithm assume loading A, loading B, processing, and storing D costs 1 clock cycle each and hence Ceil ($\beta_{PI} \cdot PI_c / PI_a$) = 4.
The second component of D_{lsp} is: Ceil ($\beta_{PP} \cdot PP_c / PP_a$). The delay ratio due to the fact that the image is larger than the array size and that chunks of the image have to be processed sequentially. Current arrays have sizes from 8*8 to 128*128, though larger sizes may be assembled. A Processor Mapping Function (PMF) is used to distribute the image points over the PEs. A PMF shows in which memory plane m and in which PE at position (x,y) of the SPA the image point (i,j) is stored. The usual PMFs are labelled: full size (one to one mapping of pixel and PE), window mapping and crinkle mapping, see figure 5.2.5 for crinkle mapping and window mapping of a 2 x 2 array on an image of size 8 x 8.

1	1	1	1	2	2	2	2
1	1	1	1	2	2	2	2
1	1	1	1	2	2	2	2
1	1	1	1	2	2	2	2
3	3	3	3	4	4	4	4
3	3	3	3	4	4	4	4
3	3	3	3	4	4	4	4
3	3	3	3	4	4	4	4

1	2	1	2	1	2	1	2
3	4	3	4	3	4	3	4
1	2	1	2	1	2	1	2
3	4	3	4	3	4	3	4
1	2	1	2	1	2	1	2
3	4	3	4	3	4	3	4
1	2	1	2	1	2	1	2
3	4	3	4	3	4	3	4

Figure 5.2.5 Crinkle mapping and Window Mapping on an SPA.

With *crinkle mapping*, every PE contains a consecutive part of the image. This means that the points which are neighbours in the original image will in general not be neighbouring points in the crinkle-mapped image. Crinkle-mapped images may therefore only be processed with a neighbourhood parallelism of 1. Because of the fact that sub-sampled versions of the image are stored, crinkle mapping may be used to do multi-scale image processing or to simulate an SIMD pyramid (Teeuw 1989; Komen 1990).

For *window mapping*, the SPA is loaded with image windows (pieces) of size ($\sqrt{P} \times \sqrt{P}$), P being the number of PEs. Although every window can be processed individually, hardware or software should provide the values of the neighbours which are across the window borders. Several methods exist to solve this 'edge-problem'.

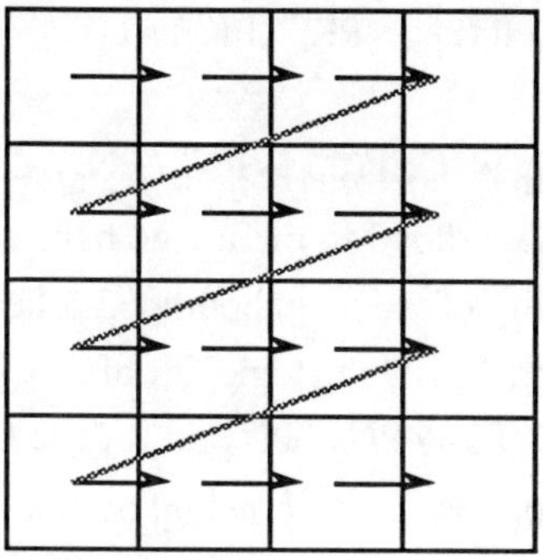

Figure 5.2.6 Scanning a small Square Processor Array over a larger image.

For instruction level processing (all image points are treated for one instruction, then for the next) the most promising methods appear to be Edge Store Scanning (ESS) or Half Scan Addressing (HSA) (Fountain 1987; Buurman 1988). SPAs are seldom equipped with special edge hardware (Fountain 1987). Figure 5.2.6 shows how a small array is scanned over a larger image. (Komen 1990) found as the range of the overhead factor : $1 \leq ß_{PP} \leq 4$
Preferably the array size should be as large as the image, and if not, certainly edge hardware should be provided to keep Ceil ($ß_{PP} \cdot PP_c / PP_a$) to PP_c / PP_a= 1, 4 and 16 for a 512^2, 256^2 and 128^2 array size respectively.

The third component of D_{lsp} is: Ceil ($ß_{PS} \cdot PS_c / PS_a$) the delay ratio due to the fact that kernels from a set have to be accessed sequentially. Due to the simple Logic Processing Unit in the SPA (no look-up table!), usually logic kernels cannot be accessed in parallel, hence each kernel takes a single clock cycle to load. Moreover for the processing of don't cares in cellular logic processing, the number of kernels doubles, yielding Ceil ($ß_{PS} \cdot PS_c / PS_a$) = 16.

The fourth component of D_{lsp} is: Ceil ($ß_{PK} \cdot PK_c / PK_a$) the delay ratio due to the fact that elements of the kernel have to be accessed sequentially. In SPAs that are designed to perform cellular logic operations Ceil ($ß_{PK} \cdot PK_c / PK_a$) is usually 1. However in SPAs designed with an emphasis on arithmetic neighbourhood processing, the cellular logic processing is usually doomed to follow the sequential kernel and neighbourhood access and hence Ceil ($ß_{PK} \cdot PK_c / PK_a$) is 2, 4 or 8, depending on the neighbourhood connectivity and parallel access possibilities. In the CLIP4 and CLIP7 the neighbourhoods can be propagated and hence accessed in parallel. We will assume Ceil ($ß_{PK} \cdot PK_c / PK_a$) =1.

The fifth component of D_{lsp} is: Ceil ($ß_{PN} \cdot PN_c / PN_a$), the delay ratio due to the fact that elements of the neighbourhood in the image have to be accessed sequentially. Usually the access possibility of the neighbourhood follows the access possibility of the kernel. In the CLIP4 and CLIP7 the logic OR of the neighbourhoods is taken and hence we will assume Ceil ($ß_{PN} \cdot PN_c / PN_a$) is 1. (The CLIP4 is the only SPA with a recursive neighbourhood. Its recursive neighbourhood access takes a number of clock cycles that depends on the data, however we will refrain from its use).

The sixth component of D_{lsp} is: Ceil ($ß_{PB} \cdot PB_c / PB_a$), the delay ratio due to the fact that bits of a pixel in the image have to be accessed sequentially. Most SPAs consist of single bit processing elements, due to the massive number of PEs. But for logic processing only a single bitplane is necessary and hence Ceil ($ß_{PB} \cdot PB_c / PB_a$) = 1.

The seventh component of D_{lsp} is: Ceil ($ß_{PO} \cdot PO_c / PO_a$). Usually SPAs are not capable to execute operations spatially in parallel, unless local autonomy is provided and the image can be separated in individually processable sub-images. The wire reduction concept makes that bit serial instruction-loading to all PEs takes place in 12 clock cycles and hence Ceil ($ß_{PO} \cdot PO_c / PO_a$) = 12.

For the skeletonization of a 512^2 image on a 512^2 PE wide SPA, $D_{im} = l \cdot D_{lsp} + D_{plt} + D_{s2i} + D_{i2d} = 256 \cdot (4 \cdot 1 \cdot 16 \cdot 1 \cdot 1 \cdot 1 \cdot 12) + 0 + 2 \cdot 8192 = 212{,}992$ clock cycles.
As t = 400,000 clock cycles (40msec) this stays within the range for real time image processing. For the skeletonization of a 512^2 image on a 256^2 and 128^2 PE wide SPA,
D_{im} = 802,816 and 3,162,112 clock cycles respectively and the real-time character is lost. In order to operate still in real-time, clock frequencies of 20Mhz and 80 Mhz respectively are necessary.
Clearly, if the number of clock cycles of the controller (accept status, prepare and load one instruction) (PO_c / PO_a) can be reduced to 1, then real-time processing is possible until SPA sizes of 128^2 with hardware edge registers or 256^2 using software scanning. Note that by proper instruction pipelining into the array, real-time

processing can be achieved. Except for data dependent actions, when the pipeline is flushed.

For the shading correction algorithm greyvalue images are used, hence 8 bitplanes should be loaded and unloaded serially yielding $D_{s2i} = D_{i2d} = 65536$ clock cycles for a 512^2 image.
The shading correction algorithm comprises mainly monadic operations, hence Ceil ($ß_{PI} \cdot PI_c / PI_a$) = 3. Moreover, CLIP4 (Duff 1982) has a working set of 32 D planes from which A, B and Carry can be loaded in a single clock pulse, loading from external image memory however costs D_{s2i} clock cycles extra as the data I/O method is used. Within this algorithm the restriction to 4 temporarily greyvalue images is not important, however for many other greyvalue algorithms like convolutions, this will lead to an overhead factor $ß_{PI} > 1$.
Ceil ($ß_{PP} \cdot PP_c / PP_a$) = 1, 4, 16 for a 512^2, 256^2, 128^2 array size respectively.
Ceil ($ß_{PS} \cdot PS_c / PS_a$) = 1, only one kernel is used.
Ceil ($ß_{PK} \cdot PK_c / PK_a$) = 4. Whereas in logic processing the entire neighbourhood can be processed in parallel, this is almost impossible for greyvalue processing. The addition unit is usually not capable to add or compare more than two values together.
Ceil ($ß_{PN} \cdot PN_c / PN_a$) = 1. (The neighbourhood can be accessed in parallel, however this is of no use here).
Ceil ($ß_{PB} \cdot PB_c / PB_a$) = 8 when using a single bit PE and a carry bitplane.
Ceil ($ß_{PO} \cdot PO_c / PO_a$) = 12, the instruction load time.
For a 21 x 21 shading correction on a 512^2 image on a 512^2 PE wide SPA, $D_{im} = l \cdot D_{lsp} + D_{plt} + D_{s2i} + D_{i2d} = (10+10+1) \cdot 1152 + 0 + 2 \cdot 65536 = 155{,}264$ clock cycles.
Real-time image processing remains possible until an array size of 256^2 using hardware edge registers.

The conclusions that can be drawn on the capability of SPA designs for real-time operation are that:
An important bottleneck in SPAs are controller cycles (PE status_out, controller operation, PE instruction in) that take longer than 1 clock cycle. This bottleneck should be cleared by proper pipelining to enable real-time processing of reasonable long image processing tasks.
If the size of the SPA is smaller than the image size, real time processing starts to become cumbersome, especially when no edge hardware is provided.

Also bottlenecks are the duration of image I/O, due to sequential data transfer over a bus with limited width, and a restricted image memory size, resulting in sequential data transfer.
The operation frequency halves if $D_{im} > t$, unless a second plane of PEs is used, and operation pipelining can be performed. However, such a solution has never been found in literature and is highly unlikely considering the number of extra PEs involved.

5.2.5 Pyramids.

A Pyramid (PYR) is a stack of two dimensional square arrays of Processing Elements. Each PE has connections to father(s) (mostly 1), in-plane neighbours (mostly 8) and a number of sons (mostly 4). Simulating an SIMD Pyramid on the CLIP4, (Teeuw 1989) concluded that simulation of a pyramidal *data-structure* in the memory of a small processor array is a possibility to increase the processing power of the array, and that the efficiency of a full size hardware pyramid is decreased by the fact that this pyramid consists of many PEs that are idling during a large part of the algorithm execution. Usually the separate planes of the pyramid are not used in an operation pipelined way and for many algorithms this is impossible due to the fact that the processing activity flows up and down the pyramid. As the advantageous features of a pyramid are only effective for a subset of the set of image processing tasks and simulated pyramids seem to be more effective than hardware pyramids, Teeuw concluded that it may be better to simulate a pyramid on a small processor array than to actually build one. Moreover, (Duff 1986) suggested that extra wire connections in the SPA plane would speed-up pyramidal processing on a flat SPA by avoiding software in-plane shifting of data.

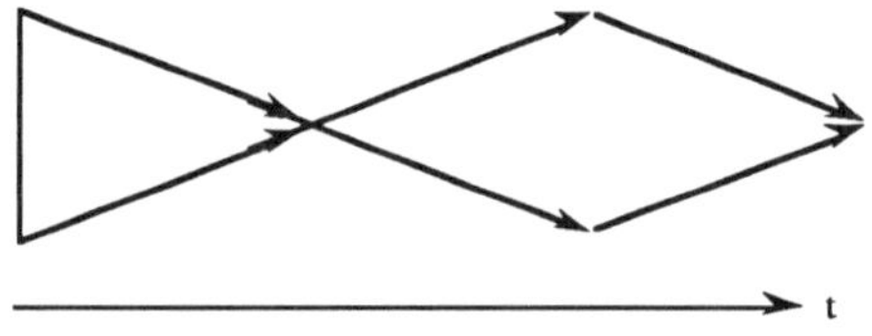

Figure 5.2.7 Operation pipelining on a full size data parallel pyramid.

A final conclusion is that a full size data parallel pyramidal architecture suffers at least the same problems as the SPA, concerning real-time processing. Operation pipelined processing on a full size data parallel pyramidal architecture while processing up and down the pyramid requires several pyramids top to bottom attached. A highly unlikely concept due to the number of involved PEs. See figure 5.2.7.

Note that the Pipelined Pyramid System (Burt and van der Wal 1987, 1990, van der Wal 1991) as discussed in section 3.5.4 is not a general purpose low-level image processing system and hence not capable of performing the two test algorithms.

5.2.6 Linear Processor Arrays.

A Linear Processor Array (LPA) has a one dimensionally connected set of *P* processing elements (PEs) to process an *N* x *N* image, see figure 5.2.8.

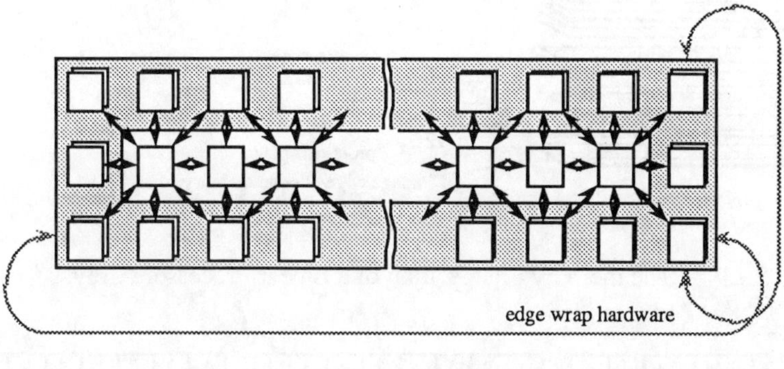

Figure 5.2.8 The data structure and PE interconnection structure of a LPA.

The LPA has a number of advantages over the SPA: A smaller number of PEs which makes application of more powerful PEs possible. If the image size *N* equals the number of available PEs, P, then every PE processes one column of the image.
The possibility to scan with the array over the image allows larger filter kernel sizes of *N* x 3 or *N* x 5. As one entire column is stored in a single PE the entire column is sequentially accessible by the PE. The memory arrangement can be the same as the SPA, albeit that synchronization with a video signal can now also be performed on a line basis instead of a frame basis. See figure 5.2.9.
If the image size *N* is smaller than the length of the LPA, a processor mapping function (PMF) determines which image point is processed by which PE. The two PMFs used for an SPA (crinkle mapping and window mapping) are also used with LPAs. The AIS-5000 (Wilson 1988) uses window mapping, and the (PICAP3

Lindskog 1988) uses crinkle mapping. A third PMF in use, is the helicoidal mapping for the SYMPATI-2 (Juvin 1988). See figure 5.2.10.

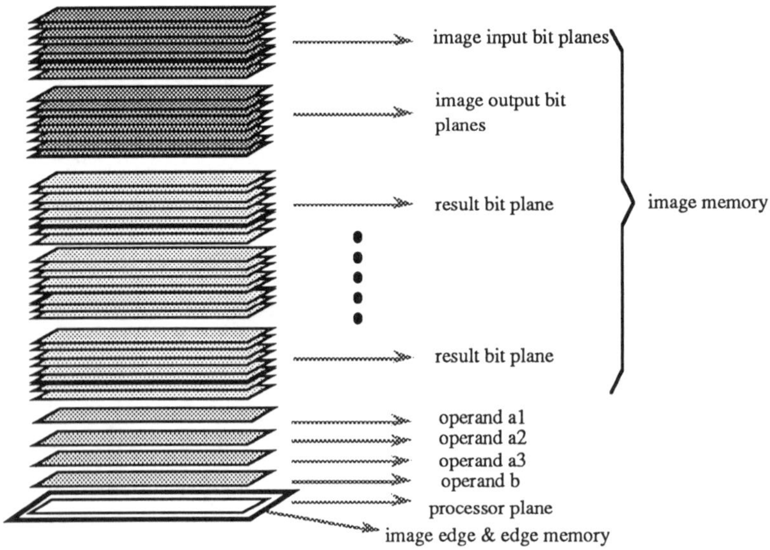

Figure 5.2.9 The memory organization of a Linear Processor Array.

1	1	2	2	3	3	4	4
1	1	2	2	3	3	4	4
1	1	2	2	3	3	4	4
1	1	2	2	3	3	4	4
1	1	2	2	3	3	4	4
1	1	2	2	3	3	4	4
1	1	2	2	3	3	4	4
1	1	2	2	3	3	4	4

1	2	3	4	1	2	3	4
1	2	3	4	1	2	3	4
1	2	3	4	1	2	3	4
1	2	3	4	1	2	3	4
1	2	3	4	1	2	3	4
1	2	3	4	1	2	3	4
1	2	3	4	1	2	3	4
1	2	3	4	1	2	3	4

1	2	3	4	1	2	3	4
2	3	4	1	2	3	4	1
3	4	1	2	3	4	1	2
4	1	2	3	4	1	2	3
1	2	3	4	1	2	3	4
2	3	4	1	2	3	4	1
3	4	1	2	3	4	1	2
4	1	2	3	4	1	2	3

Figure 5.2.10 Crinkle-, Window- and Helicoïdal mapping on a LPA.

The latter mapping makes it possible to scan the array both horizontally as well as vertically across the image. From the scanning point of view, PEs with a neighbourhood parallelism greater than one (the AIS for instance has a neighbourhood parallelism of five) will preferably use window-mapping, while other PEs may use crinkle mapping if scanning is only needed in one direction, or may use helicoidal mapping if scanning should be possible in both horizontal and vertical directions. When an LPA uses window mapping, hardware or software should provide for the values of the neighbours which are across the window borders. An LPA has a strong advantage here over an SPA, as a simple hardware scheme allows a scanning technique which does not give any overhead. A single edge memory word is sufficient. (Wilson 1989a). Figure 5.2.11 shows a small Linear Processor Array scanning over a larger image using window mapping. Due to the simple edge hardware there is no overhead factor involved and $\beta_{PP} = 1$.

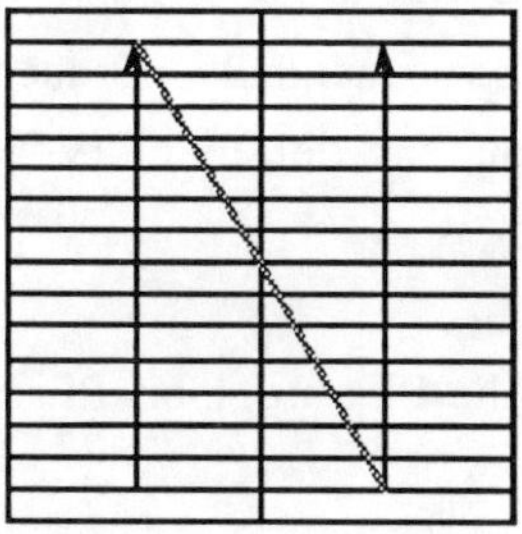

Figure 5.2.11 Scanning a small Linear Processor Array over a larger image.

If we assume a Linear Processor Array with 1 bit ALU, and an 8 input truth table module then the following calculations can be made for the skeleton algorithm:

If a real video signal is used as input, still a separate frame buffer needs to be used to create a non-interlaced video signal. This buffer can be incorporated into the memory of the LPA, it has the capability to shift in pixels from the grabbing unit and store each image line into its memory, in parallel with the processing. The same argumentation can be used for image output to a display and hence $D_{s2i} = D_{i2d} = 0$.

Ceil ($\beta_{PI} \cdot PI_c / PI_a$): Assume a system with 3 input operands; an A-plane, a B-plane and a Carry (bit)plane and one output plane (D). If A, B, C and D can all be accessed in parallel then Ceil ($\beta_{PI} \cdot PI_c / PI_a$) = 1. This is usually not the case. For the skeletonization algorithm assume loading A and B, processing, and storing D costs 1 clock cycle each and hence Ceil ($\beta_{PI} \cdot PI_c / PI_a$) = 4.

Ceil ($\beta_{PP} \cdot PP_c / PP_a$) = 512; 1024; 2048 the difference between image size and array size if the array has a length of 512; 256; 128 respectively.

Ceil ($ß_{PS} \cdot PS_c / PS_a$) the delay ratio due to the fact that kernels from a set have to be accessed sequentially. Using a look-up table, don't cares can be processed by storing all combinations in the table and hence the ratio Ceil ($ß_{PS} \cdot PS_c / PS_a$) = 1.

Ceil ($ß_{PK} \cdot PK_c / PK_a$), the delay ratio due to the fact that elements of the kernel have to be accessed sequentially = 1 due to the look-up table approach.

Ceil ($ß_{PN} \cdot PN_c / PN_a$), the delay due to the fact that elements of the neighbourhood in the image have to be accessed sequentially = 3 due to parallel access to East and West neighbourhood but non parallel memory access of North and South neighbourhood.

Ceil ($ß_{PB} \cdot PB_c / PB_a$) = 1, the PEs are 1 bit PEs and the operation is a logic operation.

Ceil ($ß_{PO} \cdot PO_c / PO_a$) = 1, Usually LPAs are not capable to execute operations spatially in parallel, unless local autonomy is provided and the image can be separated in individually processable sub-images. Due to the one dimensional structure, wiring is not so important as it is in an SPA and can be parallelized with more ease, hence Ceil ($ß_{PO} \cdot PO_c / PO_a$) = 1. Note that we make the optimistic assumption that no data depended actions are undertaken and that the instruction provision by the controller is pipelined.
For the skeletonization of a 512^2 image on a 512 PE long LPA, $D_{im} = l \cdot D_{lsp} + D_{plt} + D_{s2i} + D_{i2d} = 256 \cdot (4 \cdot 512 \cdot 1 \cdot 1 \cdot 3 \cdot 1 \cdot 1) = 1{,}572{,}864$ clock cycles.
As t = 400,000 clock cycles (40msec) this does <u>not</u> stay within the range for real time image processing. For the skeletonization of a 512^2 image on a 256 and 128 PE long LPA, D_{im} = 3,145,728 and 6,291,456 clock cycles respectively and the real-time character is still lost. For real-time skeletonization, clock frequencies of 40, 78 and 157 MHz respectively are necessary.

If more rows of PEs are introduced, operation pipelining can than be performed. See figure 5.2.12. This is e.g. possible with the Centipede LPA (Schmidt and Wilson 1989). A Pipeline of four LPAs is necessary to perform the real-time skeletonization of a 512^2 image.

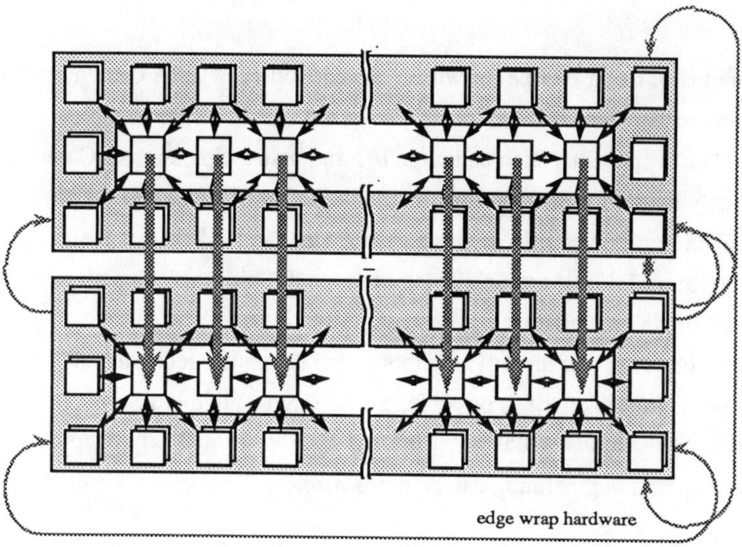

Figure 5.2.12 Scanning and pipelining with a Linear Processor Array.

For the shading correction algorithm, greyvalue images are used $D_{s2i} = D_{i2d} = 0$ clock cycles for a 512^2 image provided the image memory is large enough.

The shading correction algorithm comprises mainly monadic operations, hence Ceil
Ceil $(ß_{PI} \cdot PI_c / PI_a) = 3$.
Ceil $(ß_{PP} \cdot PP_c / PP_a) = 512; 1024; 2048$ for a 512, 256, 128 LPA size respectively.
Ceil $(ß_{PS} \cdot PS_c / PS_a) = 1$, only one kernel is used.
Ceil $(ß_{PK} \cdot PK_c / PK_a) = 4$ Whereas in logic processing the entire neighbourhood can be processed in parallel, this is almost impossible for greyvalue processing. The addition unit is usually not capable to add or compare more than two values together.
Ceil $(ß_{PN} \cdot PN_c / PN_a) = 1$. (The neighbourhood can be accessed in parallel, however this is of no use here).
Ceil $(ß_{PB} \cdot PB_c / PB_a) = 8$ when using a single bit PE and a carry bitplane.
Ceil $(ß_{PO} \cdot PO_c / PO_a) = 1$, the instruction load time.

For a 21 x 21 shading correction on a 512^2 image on a 512 PE long LPA, $D_{im} = l \cdot D_{lsp} + D_{plt} + D_{s2i} + D_{i2d} = (10+10+1) \cdot 49152 = 1{,}032{,}192$ clock cycles.
Real-time image processing is also impossible for greyvalue operations. A clock frequency of 25 Mhz would be needed for real-time processing.

The conclusions that can be drawn on the capability of LPA designs for real-time operation are that:

All possible bottlenecks such as image I/O and instruction/status I/O can be cleared more easily than with SPA designs to enable real-time processing. However, due to the reduced number of PEs in comparison with the SPA, the real-time processing capability of a LPA is marginal. This improves if 8 bit processing elements are used instead of single bit elements.

Care must be taken that the full neighbourhood is parallel accessible and that parallel logic processing of kernels is possible, e.g. through a table look-up.

The operation frequency halves at least if $D_{im} > t$, unless more rows of PEs are used, and operation pipelining can be performed.

5.2.7 Pipelines.

In a Pipeline (PL) of processors, there is only a single PE so there is no spatially or data parallelism possible. All data is sequentially processed by the PE in a dataflow, in this case a sequence of pixels from an *N* x *N* image. The PE creates is own 3 x 3 neighbourhood.

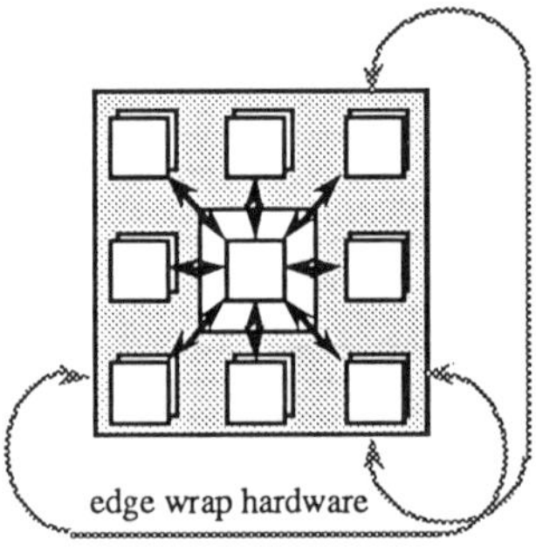

Figure 5.2.13 The data structure and PE interconnection structure of a Pipeline.

The most obvious characteristic of the PL is the operation or instruction pipelining. Each instruction or iteration of an iterative operation is executed on a successive PE in the pipeline. A hard restriction (as with the SPA) is the kernel limitation to e.g. 3 x 3. However an advantage is its flexible image size. Figure 5.2.14 shows the scanning nature of the pipeline.

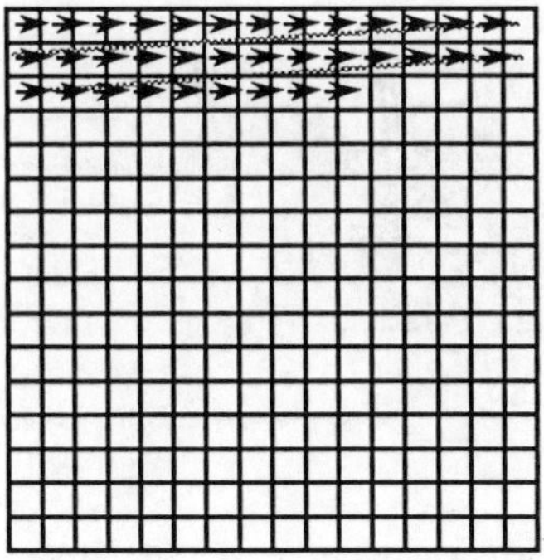

Figure 5.2.14 Scanning with a Pipeline over an Image.

If a real video signal is used as input, still a separate frame buffer needs to be used to create non-interlaced video. This buffer can be addressed to generate a sequential flow of pixels. This data is fed into the first PE, processed by it, the result is fed into the next PE and so on. The result after P PEs can be displayed or stored again in memory. The speed of the processing elements is equal for synchronization, but the PEs themselves do not have to be identical. When a pipeline incorporates different PEs it should be reconfigurable, so that an optimal path through the PEs can be made. If a PL consists of identical PEs, they should be programmable, such that a set of operations can be performed in the correct order. In this thesis we will further assume pipelines with identical programmable PEs (see figure 5.2.15).

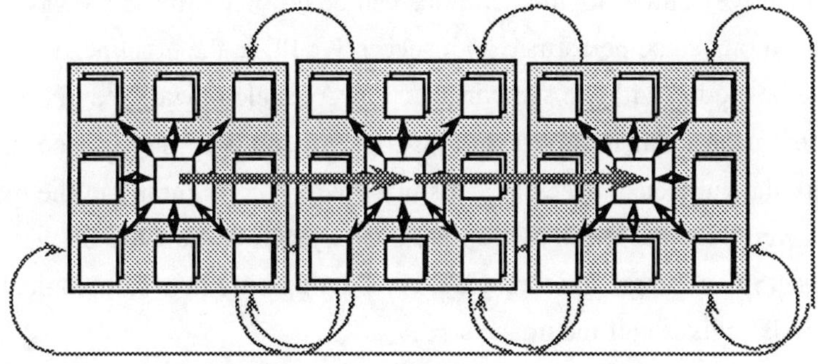

Figure 5.2.15 Pipelining with a Pipeline Processing Element.

Figure 5.2.16 shows the memory organization of the pipeline. Unlike the data parallel machines, in a pipeline the image memory is realized by the dataflow channels from which any PE can tap data and to which any PE can deflect data.

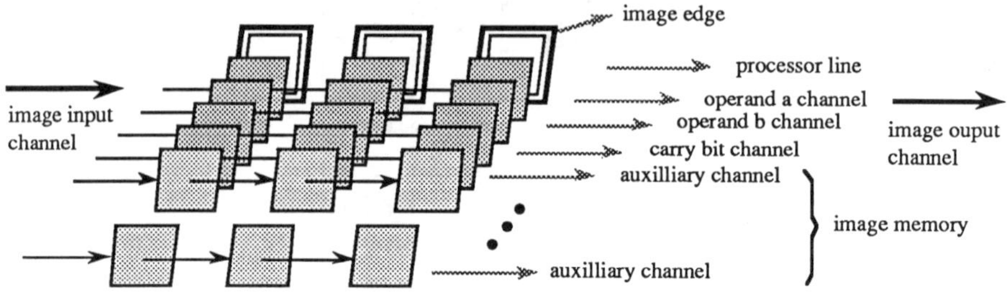

Figure 5.2.16 The memory organization of a Pipeline.

As the ratio between the amount of spatial parallelism and temporal parallelism is usually largely in favor of the temporal parallelism due to the sequential nature of most image processing operations and algorithms, the most obvious data interconnection structure is the one dimensional structure of connected PEs. However, a limited number of instructions may incidentally be processed spatially in parallel, making a second dimension with limited size sometimes profitable. However, when this is not possible, intermediate results are stored in data channels and processing takes place in a leap frog way, as is a custom in programming on sequential machines.

The conceptual idea behind a cellular logic pipeline architecture is that firstly in most image processing tasks cellular logic operations tend to appear grouped together and secondly iterative cellular logic operations can be unfolded in such a way that each separate iteration can be performed by a successive PE in the pipeline. The involved PEs should be loaded with the same instruction. As skeletonization and propagation are the most prominent iterative cellular logic operations they can be used as a measure for the number of necessary PEs in a pipeline configuration. The maximum number of iterations for these two operations is given by half the diameter of the largest object in an image. In worst case, considering only non-pathological objects (i.e. no spirals), this is half the image size N.

It is assumed that in most practical image processing tasks within one burst of cellular logic operations iterative operations such as skeletonization will hardly occur more than 4 times and the number of iterations dominate the non iterative cellular logic operations. This yields a rough estimate for the *maximum* number of PEs in such a pipeline system of 2N PEs.

Or, about 1024 PEs for a 512 x 512 image, a modest number of PEs compared to the 256,000 PEs of an equivalent 512 x 512 Processor Array.

Boundary problems on a SPA occur due to the fixed size of the main data structure, the matrix. The solution for a too small array size can be found in sequentially processing chunks of data of the size of the original array.
Boundary problems on a pipeline may occur due to a limited amount of temporal parallelism (a limited length of the pipeline). Note that lack of spatial parallelism can always be solved in a temporal (leapfrog) way, provided that enough data channels are present. This even holds for processing data words bit by bit using carry propagation through a carry channel. The problem of the limited pipeline length can be solved by processing instructions only to an amount of the length of the pipeline, using an intermediate data storage at the end of the pipeline to collect the intermediate result, recirculating the data to the input of the pipeline, loading the next chunk of instructions into the pipeline, and process the intermediate result.
Fixed boundaries are the source of the systems' inefficiency. Whereas the processor array processes all the data elements of the entire array even when many PEs contain non significant data, the pipeline has to process all data elements through to the end of the pipeline, even when the number of instructions for the task is less than the pipeline length. The rest of the pipeline should be filled with copy instructions. Provisions can be made in the form of taps onto the pipeline to allow earlier feed back of the result to the image memory. However, it is not possible to quit an unfolded iteration loop earlier than the maximum number of possible iterations that has been programmed.

If we assume a Pipeline of 512 PEs with a 1 bit Logic Unit, then the following calculations can be made for the skeleton algorithm:

If a real video signal is used as input, still a separate frame buffer needs to be used to create non-interlaced video. The PL can directly address this buffer and organize its dataflow with it. The same argumentation can be used for image output to a display buffer. As with the LPA and in contrast with the SPA this can be done in parallel with the processing but still the entire image has to be loaded into the pipeline and hence D_{s2i} = Ceil $(l/P)*N^2$. If the pipeline is too short for the number of instructions that have to be executed than the output data has to be recircled through a framebuffer. This frame recirculation has to take place Ceil (l/P) times. Hence, for the skeleton algorithm D_{s2i} = 262,144. The unloading of the image into a framebuffer takes place automatically with the processing and hence and $D_{i2d} = 0$.

Ceil ($ß_{PI} \cdot PI_c / PI_a$): As the data flows operands need not be loaded. However, the PEs create their 3 x 3 neighbourhood using shiftregisters which gives each PE an initial delay of 2 x 512 + 2 clock cycles before the first pixel of the images is available and hence Ceil ($ß_{PI} \cdot PI_c / PI_a$) = 1026.

Ceil ($ß_{PP} \cdot PP_c / PP_a$) = 1; As the ratio between data parallel datastructure and the datastructure of the PE.

Ceil ($ß_{PS} \cdot PS_c / PS_a$) = 1; All kernels including don't cares can be processed in parallel, using this type of logic unit.

Ceil ($ß_{PS} \cdot PS_c / PS_a$) = 1, using this type of logic unit.

Ceil ($ß_{PN} \cdot PN_c / PN_a$) = 1 due to the parallel accessibility of the neighbourhood.

Ceil ($ß_{PB} \cdot PB_c / PB_a$) = 1, the PEs are 1 bit Logic Units based.

The theoretical processor model is based on an execution loop:

The loop starts where the Controller instructs the Processor to take action, the Processor performs the action, reacts with status signals to the Controller, the Controller interprets the status and prepares a new instruction for the Processor. After which a new loop starts.

The factor Ceil ($ß_{PO} \cdot PO_c / PO_a$) was meant to model two things:

Firstly, the delay that the controller gives (∂); in all previous cases we have assumed that this was pipelined, and hence ∂ was 1. But secondly, the delay in the loop caused by the fact that the instruction from the Controller to the Processor apparently involved some parallel actions that could not be executed in parallel by the Processor. Consequently, the Processor sequentializes the actions by performing another Processor cycle, before reacting with status to the Controller.

In Pipelines with frame recirculation this is the case. The action ordered by the Pipeline Controller is sequentialized due to the fact that the Pipeline is not long enough. Hence $l - P$ extra Processor cycles have to be performed and consequently Ceil ($ß_{PO} \cdot PO_c$/PO) becomes Ceil ($l - P$) /P. In this case 0 as the pipeline is longer than the required number of instructions.

D_{plt} = 78 x 512 / 32 = 128 clock cycles to fully program the pipeline (CLPE) over a 32 bits bus.

For the skeletonization of a 512^2 image on a 512 PE long PL, $D_{im} = l \cdot D_{lsp} + D_{plt} + D_{s2i} + D_{i2d}$ = 1 • (1026 • 1 • 1 • 1 • 1• 1 • 0) + 128 + 262,144 = 262,272 clock cycles.

Note that l = 1 here, as it indicates the number of times that the Controller has to instruct the Processor for the operation.

For the skeletonization of a 512^2 image on a 256 PE long PL, D_{im} is also 262,272 clock cycles.

As t = 400,000 clock cycles (40msec) this stays within the range for real time image processing for both situations.

For the skeletonization of a 512^2 image on a 128 PE long PL, however, frame recirculation starts and $D_{im} = 1 \cdot ((2 \times 128 + 2) \cdot 1 \cdot 1 \cdot 1 \cdot 1 \cdot 1 \cdot 1) + 128 + (2 \times 262{,}144) = 524{,}674$. This does not stay within the range of real-time processing. A clock frequency of 13 Mhz would be needed for real-time processing.

For the shading correction algorithm Ceil ($ß_{PI} \cdot PI_c / PI_a$) also equals 1026.

Ceil ($ß_{PP} \cdot PP_c / PP_a$) = 1.

Ceil ($ß_{PS} \cdot PS_c / PS_a$) = 1, only one kernel is used.

Ceil ($ß_{PK} \cdot PK_c / PK_a$) = 4. Whereas in logic processing the entire neighbourhood can be processed in parallel, this is almost impossible for greyvalue processing. The addition unit is usually not capable to add or compare more than two values together.

Ceil ($ß_{PN} \cdot PN_c / PN_a$) = 1. (The neighbourhood can be accessed in parallel, however this is of no use here).

Ceil ($ß_{PB} \cdot PB_c / PB_a$) = 8 when using a single bit PE and a carry channel.

The number of needed PEs is: 21 x 4 x 8 = 672. Consequently, frame recirculation takes place and Ceil ($ß_{PO} \cdot PO_c / PO_a$) = 1.

D_{plt} = 128.

For a 21 x21 shading correction on a 512^2 image on a 512 PE long PL, $D_{im} = l \cdot D_{lsp} + D_{plt} + D_{s2i} + D_{i2d} = 1 \cdot (1026 \cdot 1 \cdot 1 \cdot 1 \cdot 1 \cdot 1 \cdot 1) + 128 + (2 \times 262{,}144) = 525{,}442$ clock cycles.

Note that again a higher clockfrequency of 13Mhz and frame recirculation, will restore the real-time character.

Due to the operation pipelining, at the penalty of one video frame set-up time delay, real-time image processing remains possible, however if the number of PEs is extended to 672.

The conclusions that can be drawn on the capability of PL designs for real-time operation are that:

As long as the PEs keep processing at about the video frequency, real-time processing is always possible, only the pipeline set-up time grows with longer operations. The real-time character is lost when there are not enough PEs for the operation. Frame recirculation has to be applied and the real-time character is lost.

The number of extra needed PEs is not twice as much as with the Processor Arrays, when real-time processing is a requirement.
The PL has as the advantage over the LPA that the data flows and operands and neighbourhood need not to be sequentially addressed in the image memory. This is paid by more hardware (shiftregisters). As all instructions in the PL are loaded in parallel, there is limited interaction between controller and processor possible.

5.2.8 Conclusions.

Table 5.2.1 summarizes the result of this comparison of thinning and shading correction for 512^2 images.

	PEs	Thinning (clockcycles)	Mhz	Shading correction (clockcycles)	Mhz
SPA	512^2	212,992	5.4	155,264	3.8
	256^2	802,816	20		
	128^2	3,162,112	80		
LPA	512	1,572,864	40	1,032,192	26
	256	3,145,728	78		
	128	6,291,456	157		
PL	512	262,272	6.6	524,674	13
	256	262,272	6.6		
	128	524,674	13		

Table 5.2.1 Comparison of test algorithms on the principal architectures.

For each of the investigated groups it presents the number of clock cycles and the equivalent clock frequency of the system, necessary for its real-time operation.
It shows that the Pipeline is very suitable for real-time operation, as is the full size processor array, however the latter at a cost of many more PEs, albeit simple ones.

The LPA seems not very suitable for real-time operation on first sight.
Starting with the greyvalue operation: If the LPA would have had greyvalue PEs, the number of clockcycles would have been about a factor 8 less and 3 MHZ would be sufficient for real-time operation. This cannot be accomplished for cellular logic

operations. Therefore a solution has to be found for the other bottlenecks, the sequential access of the images (factor 4) and the North and South neighbours (factor 3). As the LPA is constructed around a RAM which does not allow sequential access, shiftregisters for each image have to be used to shift in the last rows and to parallel tap the neighbourhood. The LPA performance then obtained is shown in table 5.2.2.

	PEs	Thinning (clockcycles)	Mhz	Shading correction (clockcycles)	Mhz
LPA*	512	131,072	3.3	86,016	2
	256	262,144	6.6		
	128	524,288	13		

Table 5.2.2 Performance of an LPA architecture equipped with shiftregisters.

The aim of this section was to investigate the ability of the four architectural groups for real-time processing and to review most bottlenecks and design pitfalls. Consequently we come to a final conclusion between the different design bottlenecks in SPA, LPA and PL:
The SPA is dispelled mainly because of its handling of the image as one single entity, where both sensor and display are line or pixel oriented. Moreover the two dimensional structure and the vast amount of PEs inevitably lead to compromises in the form of serial data- and instruction connections and hence in performance loss, especially when the pipelining in the controller stalls at data depended branching.

Clearly an LPA performs best if it is constituted from greyvalue PES, thus exploiting a parallelism in which it is perfectly qualified; parallel addition of whole rows of an image in a single clockcycle. If the LPA can be designed in such a way that the data flows linewise through the LPA and the total neighbourhood is parallel accessible and processable within a single clock cycle (using shiftregisters), then the LPA performs as adequate as the Pipeline for cellular logic operations.
The Pipeline performs, with less effort, as good as the LPA, for cellular logic processing as there is inherently no problem in parallel access of its memory, performed by tapping shiftregisters.

The LPA has as an advantage that, due to its controller, data dependent branching can be undertaken. In a Pipeline, data dependent branching is difficult and hence the pipeline is not flexible programmable.
The advantage of the Pipeline is that PEs can be added in quantities of one. The LPA is best to be filled with PEs in quantities of powers of 2, which leads to a second LPA to perform operation pipelining if real-time processing cannot be achieved with a single LPA.
The Pipeline is able to perform Recursive Neighbourhood Operations. This is more laborious with the LPA, that can only reach recursively the positions NW, N, NE in the downward pass or SW, S, SE in the upward pass. Image transposing and two extra passes over the image are necessary to reach recursively the E and W pixels.

5.3 A design method for special architectures.

5.3.1 Introduction.

In this section, a design method for problem decomposition and architecture specification is briefly presented. This method is largely used in the design of the CLPE.

A classical decomposition method is the separation of digital hardware in a datapath unit (the part in which data is transported and transformed) and a control unit (the part in which the transport and transformation in the datapath is controlled and ordered in time) (Blaauw 1976, Mead and Conway 1980, Mukherjee 1986, van de Goor 1989). The datapath control pair is sometimes referred to as a Functional Unit (FU). Traditionally, these Functional Units are described in a Register Transfer Language. Currently a standard design language for high level description is VHDL (IEEE 1988). See for an example of an ASIC design description using VHDL (Leung and Shanblatt 1989).

The automatic transformation from the algorithmic description via Functional Units via logic expressions to a VLSI lay-out description is often referred to as silicon compilation. The evaluation and optimization of Functional Units via logic equations to lay-out is called logic synthesis. Certain standard parts such as RAM, counters, may however be generated directly through parameterized lay-out generators. The VLSI lay-out phase that follows on the logic synthesis and lay-out generation steps comprises floor planning, placement and routing.

The high level description of the datapath often starts with some sort of data flow graph descriptions and the control unit is described using state transition diagrams or conventional (control) flow graphs as used for many years in software engineering. Data flow graphs can be decoded, optimized and scheduled into parallel and pipelined blocks (e.g. Stok 1991). Control flow graphs or state transition diagrams are transformed to Boolean equations which, minimized, often serve as input parameters for a lay-out generator (e.g. Rowen and Hennesy 1985).

However, methods and automatic tools to perform the top-level synthesis; from problem awareness via functional specification and high level decomposition until Functional Units were rare at the design time of the CLPE (1985). If existing, most decomposition methods in this fields focused and still focus on automatically transformations from an algorithm to silicon or use an input output model: "The behavior of a digital system *can be regarded* as the continual absorption of input values and the subsequent production of output values" (Krekelberg et al. 1985).

This approach is evident in the field of one dimensional signal processing, where automatically transformations from algorithm to lay-out, via an intermediate data flow analysis mostly leads to systolic pipelined systems (e.g. Dewilde et al. 1985a). However, this approach, is not necessarily applicable in every digital system design, certainly not in algorithms and designs where a class of functions operates on a single datastructure. An example of such a system is the histogram based rank filter algorithm (Huang 1979, Duin et al. 1986b, Duin et al. 1986c, Zeelen 1986) where clearly the crux of the algorithm is found in the histogram maintenance; histogram clearing, filling, updating and searching in the histogram of an n x n filter kernel to calculate a new pixel value for all pixels in the image.

The module specification method as introduced by (Parnas 1972a, Parnas 1972b) is a tool from the field of software specification that can be equally applied in the first decomposition steps of a hardware architecture (Jonker 1979, Cremers and Hibbard 1985). This method is used in the top level decomposition of the CLPE. The method starts from a set of requirements and design decisions as input to a decomposition process that ends up in modules with precisely specified interfaces. The modules that are obtained after the decomposition steps are considered to be abstract or virtual machines with its interface set specifying the module's operation. The most evident application is the incarnation of modules as abstract datatypes (e.g. Liskov 1974, Guttag 1977). However, the modules may equally be designed control oriented, data oriented, or oriented on algebraic transformations.

The decomposition process yields a hierarchical structure of modules, where modules higher in the hierarchy use modules lower in the hierarchy for their functioning and partially inherit their features. The hierarchy can be considered as a initiative hierarchy. The initiative to interaction always stems from the modules higher in the hierarchy. The decomposition process itself is a stepwise refinement method, usually not more than two levels deep and mostly only one level deep. Hence, the method is perfectly suitable for top-level decomposition and not for detailed descriptions of functionality.

The most evident differences between software and hardware is the parallelism down until the lowest level that is apparent in hardware and not necessarily in software. This implies that all modules intrinsically operate in parallel and moreover that all interfaces of the module can be used in parallel. Hence, provisions should be made to let the interfaces operate in mutual exclusion; equivalent to the monitor concept in

software (Hoare 1974). As modules are self contained and preferably stand alone testable the module interfaces often need to operate even asynchronously.
The module specification method has found its succession in the object oriented paradigm as found in e.g. the programming languages Smalltalk-80 (Goldberg and Robson 1983) and C++ (Stroustrup 1985). This object oriented approach is adopted in current high level design systems (de Lange 1991).

Further steps of the design method of the CLPE were:
Implementation of the modules by the classical decomposition into datapath and control. The datapath was described in a block diagram whereas the larger blocks in the datapath were labelled Functional Units. The finest details in the datapath are registers, counters, RAM etc. and usually Functional Blocks comprise these elements. The control was described with sets of loosely coupled Asynchronous State Machines (ASMs) transformed in a later stage to Synchronous State Machines. A module may contain more than one ASM. Each ASM with accompanying logic is also referred to as a Functional Unit. The choice for ASM modelling instead of SSM modelling is firstly a consequence of the chosen asynchronous nature of the module interfaces and secondly is the inherently timing specification of the SSM mostly superfluous on the level of description on which it is applied. Timing can always be modelled in ASMs.

As a brief resumé:
The main steps from idea to VLSI lay-out description are the following:

- Solving the kernel problem (here the mask set approach)
- Functional specification.
- High level decomposition into modules.
- Datapath and control description per module.
- Logic synthesis.
- Lay-out generation.

5.3.2 Aspects of module specifications.

Figure 5.3.1 shows the result of a first decomposition. Each module indicates a virtual machine that is capable of operating independently from and in parallel with the other modules. Each module hides certain system requirements or design decisions.

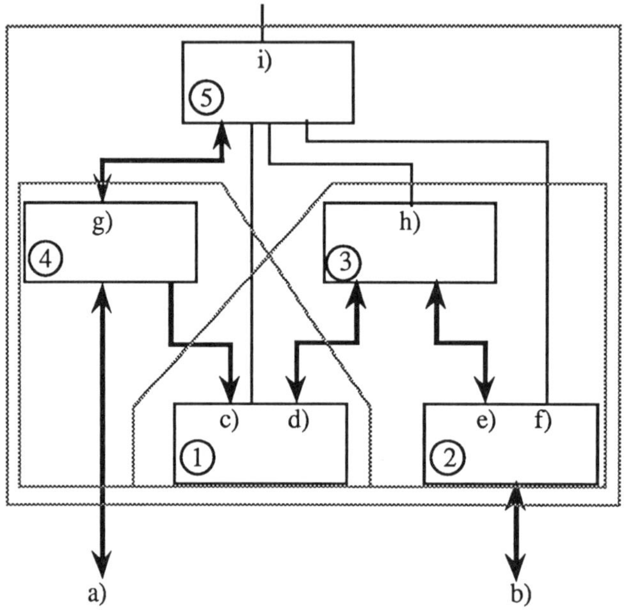

Figure 5.3.1 Main hierarchy in the CLPE.

The hierarchy in the figure is a use or initiative hierarchy. Data, control and status flow may go in any direction, only the modules higher in the hierarchy (higher in the figure) have the initiative on the communication to the modules lower in the hierarchy, the modules it uses. Thin lines indicate only control and statusflow, thick lines indicate dataflow possibly accompanied by control and status flow. Only in the data flow lines the direction of the flow is indicated, to make it possible to detect situations in which a form of mutual exclusion should be applied. This is usually where two separate datastreams enter the module. Note that modules higher in the hierarchy "use" the modules lower in the hierarchy for their operation. Consequently a precise specification of the interfaces of a module is only possible for the interfaces on top of each module. E.g. interface d) is described in module (1) and interface e) is described in module (2) but the function of d) and e) can never be described in connection with module (3). Interface d) is a service of module (1), interface e) is a service of module (2) and module (3) uses both. Consequently the precise description of each layer higher in the hierarchy specifies an increasing functionality, since it incorporates the descriptions lower in the hierarchy. The description of the top layer yields hence the description of the whole system. On the intermediate hierarchical level module 3 uses modules 1 and 2 and constitutes a virtual machine on its own. The same holds for module (4) which uses modules 1 and 2. On the top level module (5) uses modules (4) and (3) and constitutes the real machine.

5.3.3 Second decomposition in datapath and control.

Whereas the first decomposition separates the function into loosely coupled virtual machines, the second decomposition is the classic separation of the modules into tightly coupled datapath elements and a supervising control structure. Figure 5.3.2 shows the generic internal structure of a decomposed module. Both datapath unit and control unit of the module may have interfaces with modules higher as well as lower in the module hierarchy of the first decomposition; in the figure referenced by a {(module number)_interface} label. The amount of datapath and control in a module may differ from module to module. The observation is made that the larger part of modules lower in the module hierarchy usually comprise datapath and a rudimentary amount of control, e.g. only the clock. The larger part of modules in the top of the hierarchy usually comprise of control and a rudimentary amount of datapath.

The datapath.

The datapath consists of combinatorial logic blocks (L) or blocks comprising combinatorial logic and pipeline registers (L+R). Note that for proper operation of the datapath the delay from pipeline register to pipeline register should be lower than the clock period of the system. The control signals from the control to the datapath, the control vector, mostly involve register enable signals or selection signals for combinatorial logic. The status signals from the datapath to the control, the statusvector, involve special events of the datapath such as "counter zero" signals, etc.

The modules can be simulated in any normal sequential computer language using the procedure of alternately calling the control unit and the datapath unit (Blaauw 1976): Each simulation cycle, the control unit is called using the control and status signals from other modules and the status vector from the datapath. The control unit produces a control vector for the datapath as well as control and status signals to the other modules. Then the datapath is called using the data input from other modules and the control vector of the control unit, producing data output to other modules and a status vector to the control unit.

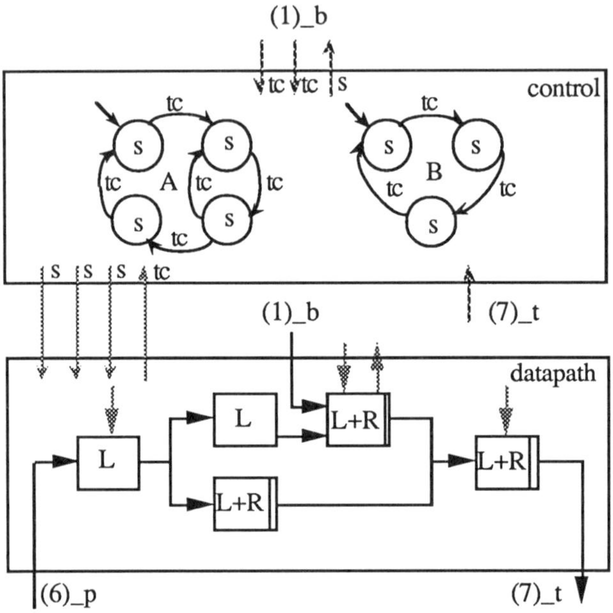

Figure 5.3.2 Module separation in datapath and control unit.

The control unit using Asynchronous Finite State Machines.

Control units are commonly modelled using Synchronous finite State Machines (SSMs). A method is proposed here to specify the control unit firstly with one or more co-operating asynchronous finite statemachines (ASMs) which are converted into synchronous statemachines in a second step. This yields a control unit that can be approach asynchronously from the outside (other modules) but operates internally and to the datapath synchronously. By keeping the interface between modules asynchronously, the interdependence of the modules remains low and makes the total system more testable. Input and output of the control unit interacts both with other modules higher and lower in the hierarchy and with the datapath (see figure 5.3.3). Loosely coupled ASMs are frequently used in specification of interfaces, e.g. the IEEE-488 interface (IEEE-1975) and in protocol specification in the field of tele- and data communication. (van Eijk et al. 1989). The benefit of loosely coupled ASMs in contrast with other control specification methods such as Petri nets (Peterson 1981) is its inherent distributed nature. It allows a uniform specification method, yet distributed over all modules possible. Moreover, by splitting up one ASMs to the smallest possible set of coupled ASMs, each ASM only performing a single function, the total number of states reduces drastically. When merging ASMs to one single ASM the number of states of the resulting ASM is the *product* of the states of the merged ASMs.

The model used for ASMs is given by the following rules, based on (Vissers 1977):

- Outputs of an ASM are based on its internal state (S) and inputs to an ASM are transition conditions (tc) that allow the ASM to make transition from its current state to its next state.
- It is assumed that the transition of an ASM takes place directly after the transition condition becomes valid and the transition itself takes an infinite short time. The ASM remains in the current state as long as the transition condition is invalid.
- Each ASM has an initial state. On a reset transition condition the ASM goes from any state to this initial state.
- Each ASM is cyclic. Through normal (non reset) transition conditions the ASM will finally transit again into its initial state.

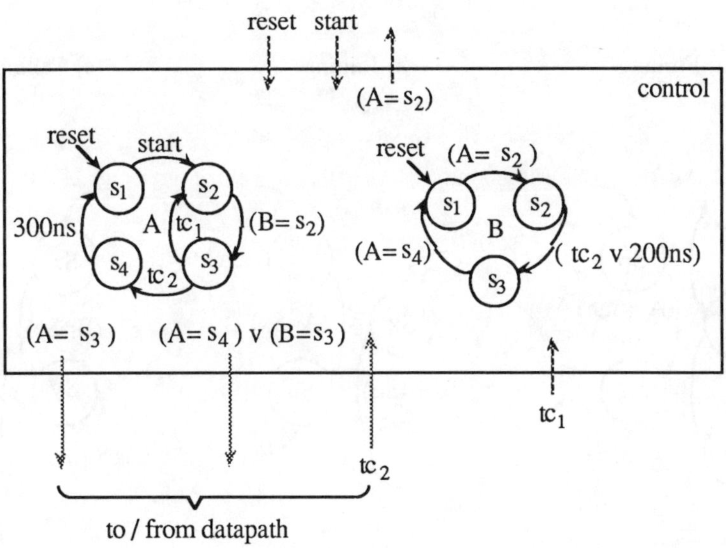

Figure 5.3.3 Specification of a control unit with co-operating ASMs.

- Transition conditions may contain any logic equation and may include timing specifications. A time-out is obtained by taking the logic OR of a normal transition condition and a time-out time. A delay is obtained by taking the logic AND of a normal transition condition and a delay-time.

- An ASM (A) is coupled to another ASM (B) by using the state condition of ASM (B) in one of its transition conditions. The coupling types between ASMs that can thus be obtained are (see figure 5.3.4):

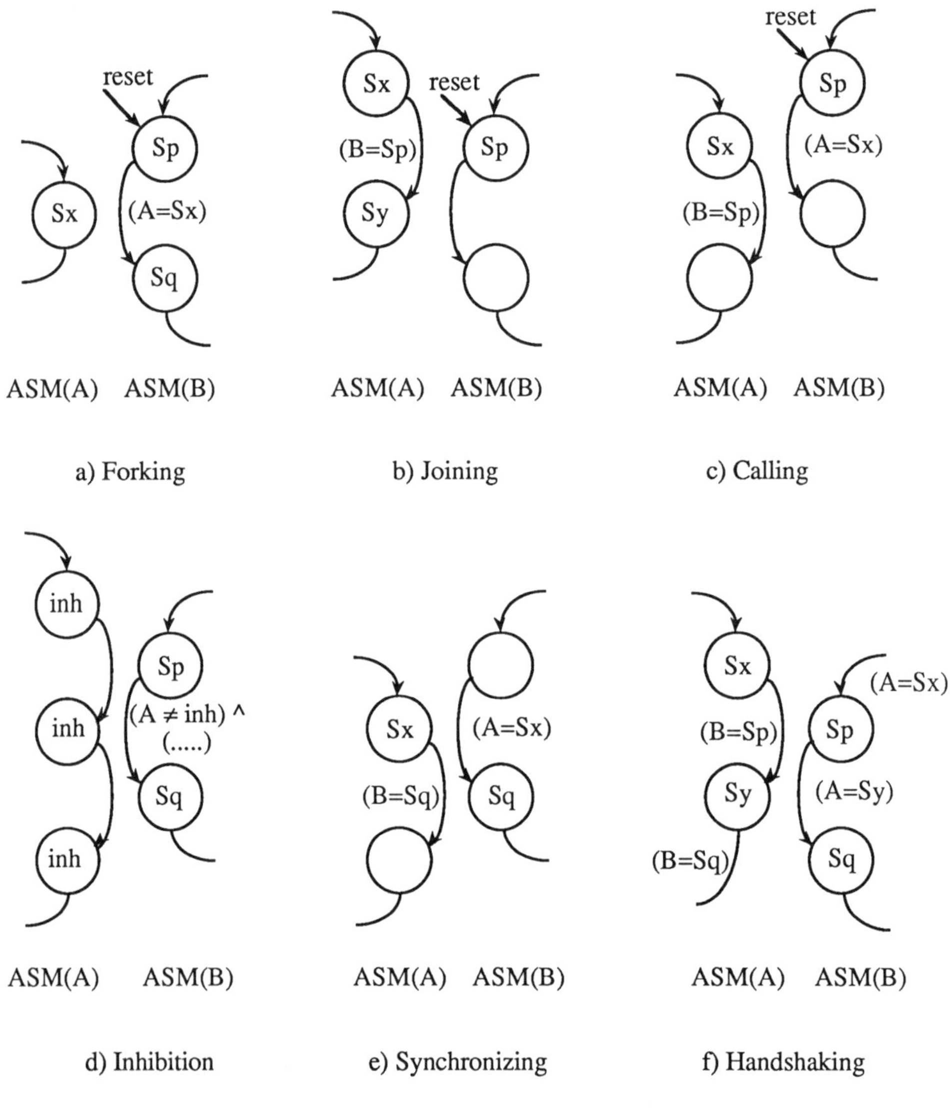

Figure 5.3.4 Possible co-operation forms between two Finite State Machines.

- ASM (A) forks ASM(B): ASM(B) is started from a state (e.g. the initial) by ASM (A) that enters a state which is a start transition condition for ASM (B). ASM(A) and ASM (B) operate further fully in parallel.
- ASM (B) joins ASM(A): ASM (A) finishes a subcycle and waits for a certain state (e.g. the initial state) of ASM(B) to continue its operation.
- ASM (A) calls ASM(B): ASM (A) starts ASM (B) and waits for the transition of ASM (B) to its initial state again. (The fork and join nodes of ASM (A) coincide with eachother).
- ASM (A) freezes ASM(B): ASM (A) starts a subcycle in which each node is a transition condition for ASM (B) that inhibits its progress.
- ASM (A) synchronizes with ASM (B): ASM (A) synchronizes with ASM (B) by forking ASM(B) and waiting for the next state of ASM (B).
- ASM (A) handshakes with ASM (B): ASM (A) handshakes with ASM (B) when ASM (A) synchronizes with ASM(B) and ASM (B) synchronizes with ASM (A).

The control unit using Synchronous State Machines.

For actual realization of the control unit in hardware and realization of absolute timing in the control module the ASMs are now transferred into Moore type SSMs. The transition of a SSM is assumed to take place on the first clock pulse after the transition condition becomes valid and the transition itself takes half a clock period, hence traversing one state of an SSM takes at least one clock period. With the control input from other modules and the statusvector from the datapath, the SSMs are updated each clock period yielding a control vector to the datapath and the other modules to be used in the next clockperiode.

Asynchronous statemachines can be made synchronously using the following rules:

- It is assumed that each synchronous statemachines is implemented using a PLA with D-type registers.
- To gain speed the states of the statemachines should be uncoded, state decoding to yield control unit output signals should be avoided. The outputs of the control unit should directly relate to bits of the state code of the statemachines.
- Inter module connections may have an asynchronous nature. This may especially be true if it concerns IC pin input signals and non clocking buffers are used. Using a clocking input buffer brings the signal under the clock discipline but may

cause a delay of one clock pulse in the reaction of the statemachine on this input signal. If delay is undesirable the asynchronous signal can be used directly in the synchronous statemachine if the state codes of present state and next state between the transition condition that includes the asynchronous signal differ only one bit in their statecode. To enhance testability it is preferable, if possible, to convert all transitions based on module input signals to single bit transitions.

- Extra states without transition conditions may be added to model a delay of a modest number of clockpulses. For more than a modest number of clockpulses a counter circuit external to the PLA should be modelled.
- Extra states without transition conditions may be inserted if this is required for a correct state assignment and if it does not harm the correct operation of the control unit.
- To obtain a correct state assignment it is allowed to assert control unit outputs on states that differ from the specified states if this does not harm the correct operation of the control unit.

The control unit specification is elucidated using a simple data acquisition (DAQ) module (see figure 5.3.5).

The DAQ module is reset by the external *reset* signal and starts AD converting on the high assertion of the *sample* signal. The 12 bits AD conversion continues until the *sample* signal is asserted low, however the last AD cycle is finished. The 12 bit values are stored in an 8 bit RAM. The AD converter starts with the signal *adst* (ad_start) and signals its completion to the control unit with the signal *eoc* (end_of_conversion). The RAM has a tristate data interface. The high byte is transferred on the signal *csh* and the low byte on *csl*. The data is clocked into the RAM on the assertion of the signal *ds*. The address counter is cleared by the signal *clr* and incremented by *inc*.

Figure 5.3.5 shows the control unit with a single ASM. Note that the output signals indicated in the states are asserted high as long as the ASM remains in that state, otherwise these signals are asserted low. The analysis on synchronous and asynchronous signals is performed in figure 5.3.5 using the / to indicate a required single bit state transition and the // if a multiple bit state transition is allowed.

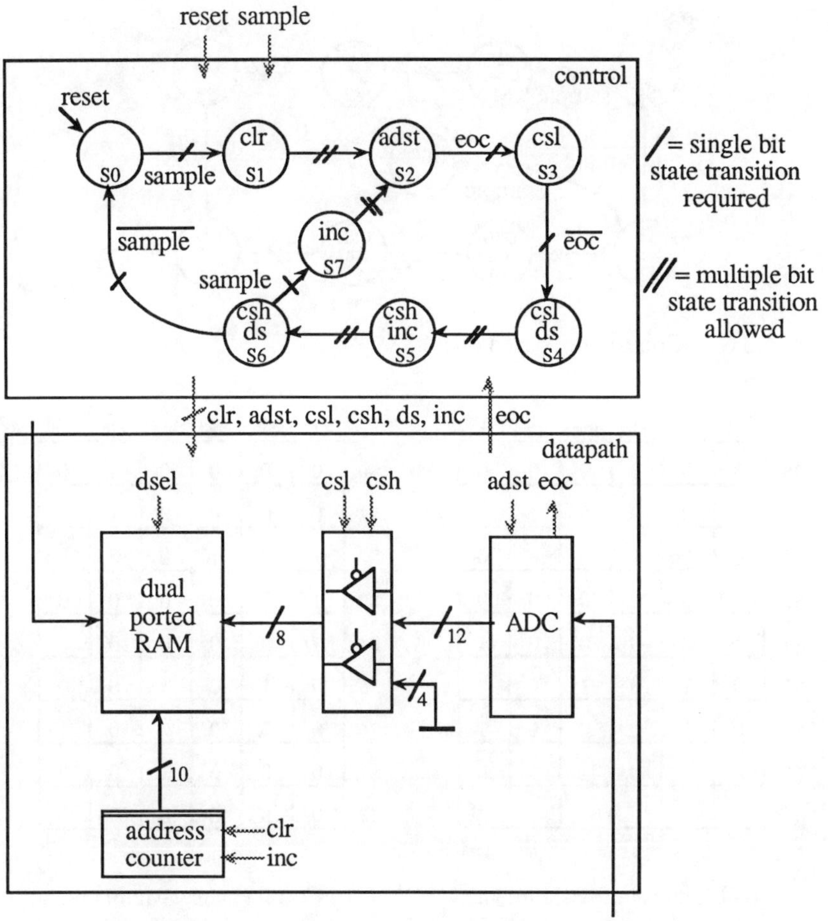

Figure 5.3.5 Datapath unit and control unit with ASM

The ASM of figure 5.3.5 is now transferred to a SSM. Table 5.3.1a shows a state assignment based on direct usage of the output signals as state bits. The transition problems to be solved are:

- S_2-S_3 a multiple bit transition, should be single. Solved by adding *csl* to the state which does not harm the operation (see table 5.3.1b and figure 5.3.6).
- S_6-S_0 and S_6-S_7 multiple bit transitions, should be single. Solved by adding an extra state S_8 to the SSM and adding *clr* to the state S_0 which does not harm the operation. Note that the penalty here is one clockpulse delay after each AD conversion.

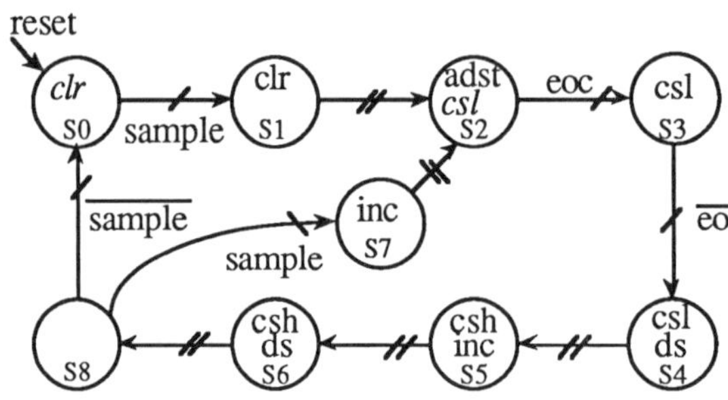

Figure 5.3.6 Control unit based on a SSM.

	csl	csh	adst	inc	clr	ds
S0	0	0	0	0	0	0
S1	0	0	0	0	1	0
S2	0	0	1	0	0	0
S3	1	0	0	0	0	0
S4	1	0	0	0	0	1
S5	0	1	0	1	0	0
S6	0	1	0	0	0	1
S7	0	0	0	1	0	0

a) Original state assignment

	csl	csh	adst	inc	clr	ds
S0	0	0	0	0	1	0
S1	0	0	0	0	1	0
S2	1	0	1	0	0	0
S3	1	0	0	0	0	0
S4	1	0	0	0	0	1
S5	0	1	0	1	0	0
S6	0	1	0	0	0	1
S7	0	0	0	1	0	0
S8	0	0	0	0	0	0

b) Correct state assignment.

Table 5.3.1 State assignments based on direct use of the input signals.

The resulting SSM can be used to feed a PLA lay-out generator.

5.3.4 Simulation.

System specification and simulation can be performed on various levels. Preferably all specifications should be executable in a simulation system and the same simulation system should be used throughout all design steps. In this case, systems with not yet implemented or realized modules or units can be simulated together with already detailed modules on different levels, and the system can be simulated within its final environment, e.g. incorporated in user software. Most important is that the total

system simulator model can be kept consistent on all levels throughout the design project and during a long part of the systems live to allow correct system updates.
The simulation system can be embedded in a suitable high level computer language such as LISP (Blackman et al. 1985) or APL (Blaauw 1976, Brenner 1984) to use the expression power of the language and at the same time enable a fluently integration of synthesis tools (Brayton 1987). A proper simulation is set up as a datapath simulator and a control unit simulator. Each clockpulse, the control unit is called, yielding a control vector for the datapath, followed by a datapath call, yielding a statusvector for the control unit.

5.3.5 Conclusions.

The top level description of the CLPE was set-up in APS (de Graaf and van Leuken 1984), a variant of the register transfer language ISPS (Barbacci 1987). In parallel some details, mainly concerning the mask processing, have been simulated in APL. In a fully different environment the low-level NMOS and later on CMOS description and simulation was performed in SLS (Dewilde et al. 1985b). This was a severe drawback for the system design as it was not possible to construct and maintain a mixed high level and low level system description. However, inevitably due to the state of the art at the start of the CLPE project (1985).
The used design method proved to be very powerful as it yields a clear hardware structure, which eases the testing of the device.
APL (Iverson 1962, 1980) remains a powerful tool for hardware description. The PE for image processing as described in chapter 6 was simulated in APL. The language has many powerful mathematical language constructs, as well as a capability to describe logic on the lowest level. This allows a smooth construction of architectures as implementation details can be temporarily omitted, by representing them by high level constructs. Moreover, due to its very high level it allows seamless interaction with the user environment of the application. For example in image processing, image memory, test image generation and image display tools are easily programmed and connected to the architecture. However, a drawback is that it lacks coupling with CMOS lay-out description systems.

5.4 The architecture of a Cellular Logic Processing Element.

5.4.1 Design aspects of the CLPE.

In this section the design of an architecture for a real-time low level image processing Pipeline is discussed. This architecture implemented as a CMOS VLSI circuit for Cellular Logic Processing is a device that is able both to operate as a single processor and as a Processing Element for a real-time binary image processing pipeline.

The design served many goals:

- To provide a basis for the comparison of massive data parallel SIMD architectures and instruction parallel pipelines.
- To obtain experience in VLSI design as inevitable vehicle for future architectures.
- To support the research on mathematical morphology in multi-dimensional images.
- And inevitably, to function as a test object for VLSI design tools under development.

The design of the Cellular Logic Pipeline Element was clearly based on the design of the Cellular Logic Table Module (LOGT) of the DIP-1 (Gerritsen and Aardema 1981, Gerritsen 1982) and its stand-alone VME board embodiment (Boekamp et al. 1985). Both designs are single processors able to process a flow of binary pixels originated from a raster scan through an image. For this purpose, shiftregisters are used to create the 3 x 3 neighbourhood while the processing itself was based on look-up tables in RAM or ROM respectively.
The CLIP4 design (Duff 1984) influenced the CLPE in the sense that the CLIP4 can be considered as the archetype of a massively data parallel processor array for cellular logic image processing. The CLPE was sought to be its counterpart: A massively instruction parallel machine for cellular logic image processing. This aim was inspired by research on the comparison between processor array architectures and pipeline architectures for low level image processing (Duin and Jonker 1986a) which eventually led to the thesis of (Komen 1990a). As a consequence of this aim the properties of both processing elements and the properties of both systems were sought to become as similar as possible. However, with an open eye for the pitfalls of past machines.

As in the Clip4 the CLPE should process single bits of data. Greyvalue processing in the Clip4 takes place sequentially, bitplane by bitplane, made possible by a carry bit register in the PEs. In case of the CLPE a carry provision should be made on each PE to allow greyvalue processing. Greyvalue processing can now either be performed by cascading more single bit pipelines in the second dimension, usable anyway for spatial parallelism in binary processing, by using leapfrogging, or by using both. This subject will be discussed in more detail in section 5.5.

Using the theory of the sections 5.2 and 5.3, a special Cellular Logic Processing Element (CLPE) was designed (Jonker and Duin 1985, Kraaijveld et al. 1986, Jonker et al. 1988b). This device is a pipeline processing element, programmable for cellular logic operations based on sets of 3 x 3 structuring elements or masks. Up to 30 masks can be downloaded during operation, while the masks can be grouped to a set by providing them with an identical label. The CLPE has been designed using a precise decomposition and simulation method and has been realized in the Philips C5TH (N-well, Single Layer Metal, 3 micron CMOS) process as a 40 pins device (Schot 1988). The clock frequency was aimed at 10 Mhz resulting in a data-throughput of less than 100ns. per pixel. The chip lay-out contains about 35.000 transistors in over 200 different cells in a full-custom VLSI technique.

5.4.2 Requirements and restrictions.

The considerations in sections 5.2 and 5.3 led to a list of requirements and restrictions that have been used for the design of the CLPE. This list is summarized below:

The CLPE should be able ....

a) to function as a stand alone processor, able to address, read and write into image memory.
b) to process images of at least 512 x 512 pixels for image processing purposes.
c) to process images up to 16K x 16K pixels for VLSI mask checking and Printed Circuit Board checking. An important industrial application field.
d) to perform all known cellular logic operations in a 3 x 3 neighbourhood including recursive neighbourhood operations.
e) to perform general logic convolution in a 3 x 3 neighbourhood, i.e. all unknown cellular logic operations in a 3 x 3 neighbourhood.

f) to perform binary rank filtering in a 3 x 3 neighbourhood. A still important set of cellular logic operations, e.g for spot noise reduction.
g) to iterate an operation for a programmable number of times or until idempotence occurs at some iteration step. In stand-alone use of the CLPE and iterative cellular logic operations.
h) to perform all possible monadic and dyadic binary operations as preprocessing step as well as post processing step to the cellular logic operation or cellular logic iteration. Binary post processing in iteration steps of recursive neighbourhood operations makes the propagation and anchor skeleton operation possible.
i) to return the number of iterations to the user. Often used as a rough measure for the radius of the largest object in the image.
j) to return the number of foreground pixels to the user. Often used as a measure for the remaining number of objects in the image.
k) to function as a processing element in a processor pipeline. Configuration in a pipeline makes processing of a sequence of cellular logic operations as well as unfolding of iteration loops of iterative operations possible.
l) to provide an image bypass channel within the chip in addition to the image process channel. To implement spatially parallelism or to create a possibility to combine intermediate results derived earlier in the instruction sequence.
m) to iterate with upward as well as downward passes through the image to speed up recursive neighbourhood operations.
n) to be started and reset by a host and to interrupt this host when processing is finished.
o) to be used in a real-time environment. The frequency was aimed at 15Mhz, however aiming at 10Mhz seemed more realistic with the C5TH process in mind. During video line and video frame sync times the invalidity of the data should be signalled to the processing element.
p) to be interrogated, loaded and instructed during process time.
q) to have provisions for grey value operations. This requirement was dropped during the design as too ambitious for that moment. See also section X.
r) the correct functional behavior of the VLSI circuit should be made testable through its bonding pads (pins).

The CLPE should be restricted to....
s) a 40 pins Dual In Line case. Allowing a high packing density on printed circuit boards.

t) a die-size of about 10 x 10 mm. To allow packing in a DIL case.

u) such a design that as less additional chips as possible are necessary for its correct functioning.

In the following sections the decomposition of the CLPE into modules and the decomposition into datapath and control using the methods explained in section 5.3 can be performed using this list.

5.4.3 Decomposition in modules.

Module functionality.

Clearly one of the first modules that can be distinguished is a module that performs the actual cellular logic processing function in the 3x3 neighbourhood. This module implements requirements d) *"to perform all known cellular logic operations in a 3 x 3 neighbourhood including recursive neighbourhood operations"*, e) *"to perform general logic convolution in a 3 x 3 neighbourhood, i.e. all unknown cellular logic operations in a 3 x 3 neighbourhood"* and f) *"to perform binary rank filtering in a 3 x 3 neighbourhood"*. Knowing already the solution to implement the actual processing using mask sets in a Writable Logic Array, the module will be called Mask Module.

The second module interfaces the chip with the outside world; the Data Interface Module. This module implements requirements a) *"to function as a stand alone processor, able to address, read and write into image memory "*, m) *"to iterate with upward as well as downward passes through the image"*, k) *"to function as a processing element in a processor pipeline "*, o) *"to be used in a real-time environment. "*.

A third module arranges the processing of images in a flow, generates the 3 x 3 normal neighbourhood, the recursive neighbourhood and performs the binary pre and post processing on the image flows. This module implements requirements b) *"to process images of at least 512 x 512 pixels "*, c) *"to process images up to 16K x 16K"* , h) *"to perform all possible monadic and dyadic binary operations as preprocessing step as well as post processing step"*, g) *"to iterate an operation for a programmable number of times or idempotence "*, l) *" to provide an image bypass channel "*, i) *"to return the number of iterations"*, j) *" to return the number of foreground pixels "*. Requirement q) *"to have provisions for grey value operations"*

although not implemented in the design of the CLPE chip itself belongs to this module as carry data is just another image flow that need be processed.
This module will be called process module for obvious reasons. The module will use the Mask Module and the Data Interface Module for its functioning.

The remaining requirements: n) "*to be started and reset by a host and to interrupt this host when processing is finished*" and p) "*to be interrogated, loaded and instructed during process time*" point in the direction of an instruction interface module. Requirement r) "*the correct functional behavior should be made testable through its pins*" points in the direction of a programmable re-definition of the pins of the chip, such that the pins give access to internal circuits in the chip for test purposes.
Moreover restriction s) "*a 40 pins Dual In Line case*" gives a limitation to the number of pins. Knowing the amount of bits to control the process module and the amount of status bits that should be provided back to the host, a decision was made to use a serial interface to load instructions and write status to the host.
Consequently an Instruction Interface Module implements requirement p) "*to be interrogated, loaded and instructed during process time*" and restriction s) "*a 40 pins Dual In Line case*" by using a serial communication structure.
A Master Module implements requirements n) "*to be started and reset by a host and to interrupt this host when processing is finished*" and r) "*the correct functional behavior should be made testable through its pins*" .
Figure 5.4.1 shows the above decomposed structure.

Module interfaces

The Mask Module (1) will have two data interfaces specified as:

1.1)	result <- associate(neighbourhood);	give the result associated with this neighbourhood.
1.2)	down_load(mask_opcode);	download one mask-opcode combination.

And one control interface specified as:

1.3)	select_clop(opcode);	select with the operation code which cellular logic operation should be used.

The module uses no other modules for its functioning.

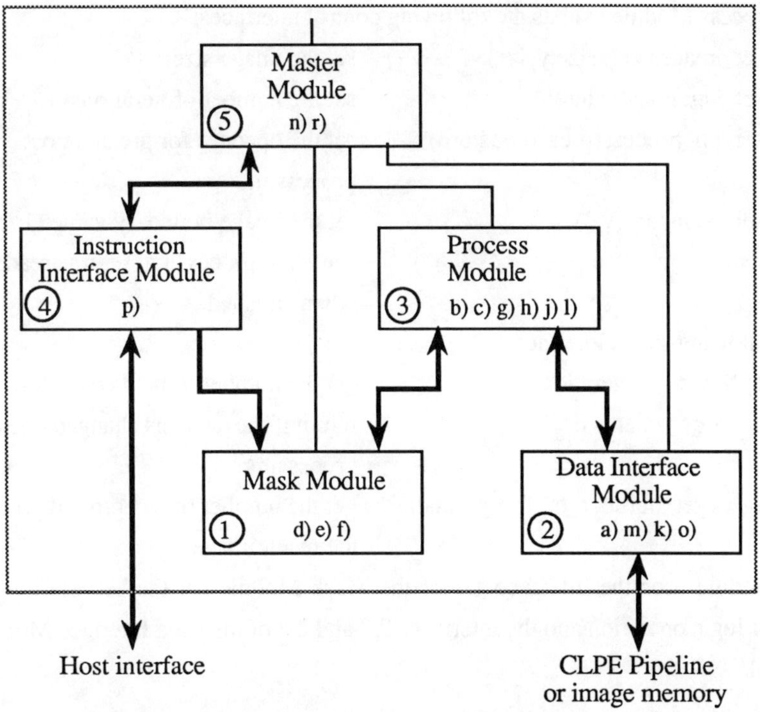

Figure 5.4.1 Main hierarchy in the CLPE.

The Data_Interface Module (2) will have the following control interface:

2.1) set_channel(p_or_m, chan, address); set up a synchronized pixel flow between either one of the pipe interfaces or one of the image memory planes and one of the internal channels.

And two data interfaces:

2.2) inflow <- image_in(chan, i_sync); read a pixel into one of the internal channels synchronously with the outside world.

2.3) outflow <- image_out(chan, o_sync); write a pixel from one of the internal channels synchronously to the outside world.

The module uses the data pins of the CLPE.

The Process Module (3) has the following control interfaces:

3.1)	set_imagesize(xsize,ysize);	set the image size.
3.2)	set_iterations(value);	set the number of iterations.
3.3)	set_pp_process(prepro,postpro);	set the opcodes for pre and post processing.
3.4)	set_boundary(val);	set the image boundary value (1 / 0).
3.5)	ready <- runclpe	starts the processing, returns ready when finished.

The following status interfaces:

3.6)	cycles <-get_iterations;	get the number of iterated cycles.
3.7)	cd <- get_changed;	return if the data has changed due to the processing.
3.8)	fg <- get_number_of_foreground;	get the number of foreground pixels of the result image.

The module uses the interface 1.1 of the Mask Module for the association of the cellular logic operation and the interfaces 2.2 and 2.3 of the Data Interface Module for image I/O.

The Instruction Interface Module (4) has one control interface:

4.1)	cmd <- get_command;	get interface command; load instruction, load wla_word, read status, run

And two data interfaces:

4.2)	instruction <- in_flow;	get an instruction from the host.
4.3)	out_flow(status);	put status to the host.

The module uses the control pins of the CLPE.

The Master Module (5) has one control interface:

5.1)	reset_clpe;	resets the clpe.
5.2)	test;	change the CLPE pin definition to the test definitions of the pins to allow access to internal circuits.

The master module uses interfaces 4.1, 4.2 and 4.3 of the Instruction Interface Module, the interface 1.3 of the Mask Module, the interface 2.1 of the Data Interface Module and all interfaces of the process module.

Figure 5.4.1 shows the result of the first decomposition for the CLPE. Each module indicates a virtual machine that is capable of operating independently from and in parallel with the other modules. Each module hides certain design decisions; requirements from the list. Thin interface lines indicate only control and status flow, thick interface lines indicate data flow possibly accompanied by control and status flow. Only in the data flow lines the direction of the flow is indicated, to make it possible to detect situations in which a form of mutual exclusion should be applied. This situation occurs in the mask module that is both used by the instruction interface module (download 1.2) and the process module (associate 1.1).
Note that modules higher in the hierarchy "use" the modules lower in the hierarchy for their operation and hence inherit their properties. Consequently the description of each layer of modules higher in the hierarchy specifies an increasing functionality, since it incorporates the descriptions lower in the hierarchy. The description of the top layer gives the description of the whole system.

An implementation reminder for the Mask Module is: "*Mutual exclusion between 1.1 and 1.2* " And the remaining remarks from the restrictions list were: t) "*a die-size of about 10 x 10 mm*". This is applicable for all modules, but the margin is in the number of masks of the Mask Module. And u) "*such a design that as less additional chips as possible are necessary* ". This is mainly applicable to the interface modules.

5.4.4 Second decomposition in datapath and control.

The mask module.

The heart of the CLPE is formed by the mask module. The module has two sub-blocks, the Writable Logic Array (WLA) and the Tally-circuit. Figure 5.4.2 shows the implementation of the Mask Module with two WLA blocks and one Tally Circuit. The Mask Module mostly consists op datapath and has hardly any control. Data path signals are black and control signals are shaded in figure 5.4.2.
In section 5.1 the WLA association function was defined as:

$$\text{Result\_Out} \leftarrow \bigvee_{\forall \text{rows}} \{\text{Invert} \neq \overline{(\text{And\_plane}^* \,(\vee.\wedge)\, \text{Nbh\_Opc}^*)}\}$$

(5.1.11)

or: Result_Out <- associate(nnbh, rnbh, mi, opcv).

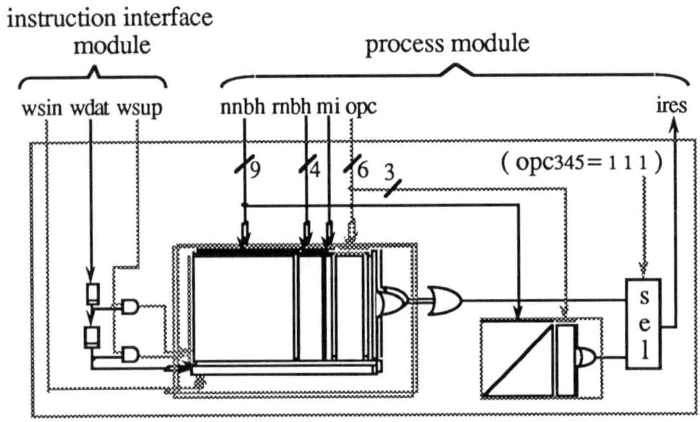

Figure 5.4.2 The mask module.

It appeared to be possible to implement a maximum of 30 WLA terms remaining within the restriction of t) "*a die-size of about 10 x 10 mm*". However speed limitations due to long signal lines made it necessary to split the WLA into two blocks of 15 terms each, yielding two functions, one for each WLA. The data interface:

1.1) result <- associate(neighbourhood) can now be implemented with:

{

(nnbh, rnbh, mi, opcv) <- (neighbourhood)

result1 <- associate1(nnbh, rnbh, mi, opcv)

result2 <- associate2(nnbh, rnbh, mi, opcv)

result3 <- tally(nbh, opcv[012])

ires <- {(opcv[345] = 1 1 1) | result3} $\vee$ {(opcv[345] $\neq$ 1 1 1) | (result1 $\vee$ result2)}

}

The pin limitation leads to serial mask data at the CLPE input. For the sake of a limited number of signal lines it was more parsimonious to incorporate the serial shift register for the input bits of a mask_opcode vector, near to the WLA itself instead of in the Instruction Interface Module. Due to the implementation of the WLA, it was not possible to hide the existence of two WLAs for the user. In the WLA mask_opcode vector that is shifted into the shift register of the WLA two extra bits have been concatenated to identify the WLA that should be loaded. Hence, this vector consists now of:

mask_opcode = (wla_id, nnbh_nd, rnbh_nd, mi_nd, opcv_nd).

with: wla_id = {0 0, 0 1, 1 0, 1 1}
for: {disable write, load in WLA1, load in WLA2, load in both}.

Figure 5.4.2 shows how the two last bits are not shifted into the shift registers of the WLA but are used to enable or disable the shift_up lines of the two WLAs. The download interface function:

1.2) down_load(mask_opcode); can be implemented with:

```
down_load(sin, sup, wdat);
{
    (wla_id, nnbh_nd, rnbh_nd, mi_nd, opcv_nd) = wdat
    if (wsin)
        {
            ∀ column = maxcolumn .. 1 shift_register1[column] <-
            shift_register1[column-1]}
            ∀ column = maxcolumn .. 1 shift_register2[column] <-
            shift_register2[column-1]}
            shift_register1[0] <- wdat[column]
            shift_register2[0] <- wdat[column]
        }
    wsup1 <- (shift_register1[0] ∧ wsup)
    wsup2 <- (shift_register1[1] ∧ wsup)
    if (wsup1)
        {
            ∀ row = maxrow .. 1 {and_plane1(row) <- and_plane1(row-1)}
            and_plane1(0) <- shift_register1
        }
    if (wsup2)
        {
            ∀ row = maxrow .. 1 {and_plane2(row) <- and_plane2(row-1)}
            and_plane2(0) <- shift_register2
        }
}
```

The mutual exclusion of the functions 1.1 (associate) and 1.2 (down_load) has been realized on the lowest level, allowing the WLA being loaded during process time. (requirement p)). The CLPE uses a 2 phase clock scheme $\{Ø_1, Ø_2\}$. At each $Ø_1$ clock cycle a download shift up action and at each $Ø_2$ clock cycle an associate action can be performed. This makes it possible that a new mask set is loaded during process time. The position of a mask does not influence its operation. However, as masks are shifted out at the top for each WLA, the user should keep a book of the positions of the masks in actual use in each WLA.
In order to show the design process a precise description was given of each of the interfaces of the Mask Module. These descriptions can be used to set up a simulation of these modules. See section 5.1. In the explanations of further modules these descriptions will be omitted.

The process module datapath.

The process module performs the main function of the CLPE. It uses the Mask Module and the Data Interface Module. Figure 5.4.3 shows the datapath of the process module. The black arrows indicate data flow, the shaded arrows indicate control flow.

As can be observed the datapath has 4 data input channels. The input selector can be used to connect any of the four sources (*s1, s2, s3, msk*) to any of its 4 input channels (*a, b, c, d*). The selection is performed by the control signals (*src1, src2, src3, msk*) from the master module. The sources (*s2, s3, msk*) can be enabled and disabled by the settings of (*src2, src3, msk*). As there should always be one source image, source *s1* cannot be disabled. On the three source inputs (*s1, s2, s3*) pair wise dyadic Boolean operations can be performed using two Boolean Function Generators.
With a Boolean function generator any monadic and dyadic Boolean function can be performed. See table 5.4.1. The output of the Mask Module (*ires*) is combined with the delayed mask image with a third Boolean function generator to allow post processing of the cellular logic result. This post processing together with the feedback of the *result* to create the recursive neighbourhood, enables the Anchored Skeletonization and Propagation operations (requirement h)). The destinations (*result, mask, original*) are passed to the Data Interface Module.

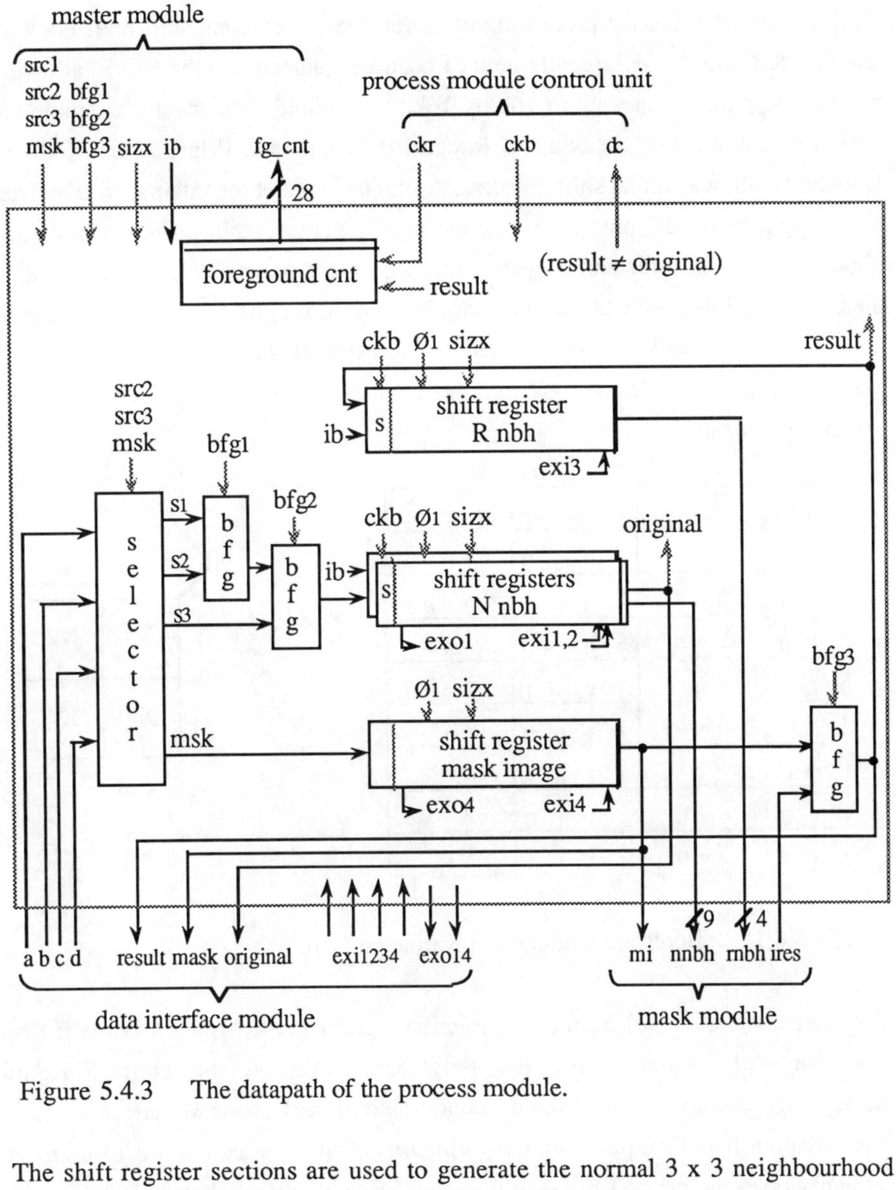

Figure 5.4.3 The datapath of the process module.

The shift register sections are used to generate the normal 3 x 3 neighbourhood (*nnbh*), the recursive neighbourhood (*rnbh*) and the delay of the mask image (*mi*) which are input to the Mask Module. The shift registers are permanently clocked by $Ø_1$. The processor itself is equipped with shift registers and a count sequence to process images of 512 bits in one direction by up to 16k pixels in the other direction. The shift registers have a length (*sizx*) that is programmable in powers of 2, with $32 \leq sizx \leq 512$ for full image as well as Region of Interest (ROI) processing purposes (requirement b)). For the processing of larger images, external shift registers with length (1K $\leq$ *sizx* $\leq$ 16K), requirement c)) can be connected to the CLPE, allowing the processing of images up to 16k by 16k. This would facilitate for example the design rule and mask checking of Integrated Circuit and Printed Circuit Board layouts. To allow external shift registers, the connections of the internal shift register are brought to the IC pins. The external shift registers replace the internal shift registers. The use of the shift registers has been made transparent. This means that the CLPE itself decides whether it can use its own shift registers, or whether it has to use external shift registers (if connected). The external shift connections are (*exi1, exi2, exi3, exi4*) and (*exo1, original, result, exo4*). Note the double function of *original* and *result.*

$x_3\ x_2\ x_1\ x_0$	operation	$x_3\ x_2\ x_1\ x_0$	operation
0 0 0 0	0	1 0 0 0	$a \wedge b$
0 0 0 1	$\bar{a} \wedge \bar{b}$	1 0 0 1	$a = b$
0 0 1 0	$a \wedge \bar{b}$	1 0 1 0	a
0 0 1 1	$\bar{b}$	1 0 1 1	$a \vee \bar{b}$
0 1 0 0	$\bar{a} \wedge b$	1 1 0 0	b
0 1 0 1	$\bar{a}$	1 1 0 1	$\bar{a} \vee b$
0 1 1 0	$a \neq b$	1 1 1 0	$a \vee b$
0 1 1 1	$\bar{a} \vee \bar{b}$	1 1 1 1	1

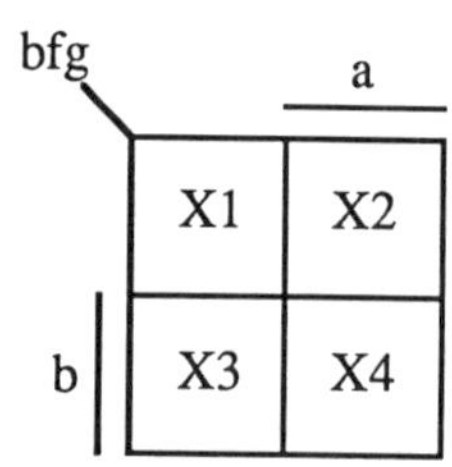

Table 5.4.1 Boolean functions and Karnaugh map.

The image boundary (*ib*) from the instruction register in the Master Module is read into the shift registers for the neighbourhoods on the control signal *clock_image_boundary* (*ckb*) from the control unit of the Process Module..

The datapath has 3 output channels, which can either be connected to memory bitplanes or to the next CLPE(s) in the pipe. The outputs are the result image, the (delayed) original image and the (delayed) mask image. Figure 5.4.5 shows a limited possibility of forking, quasi parallel operation and joining of results using CLPEs. This is possible by the feed trough of the original and mask image, thus satisfying requirement l). Note that the fourth CLPE input channel remains free in pipeline operations for carry connections between CLPEs, hence partly satisfying requirement q). See also section 5.2.

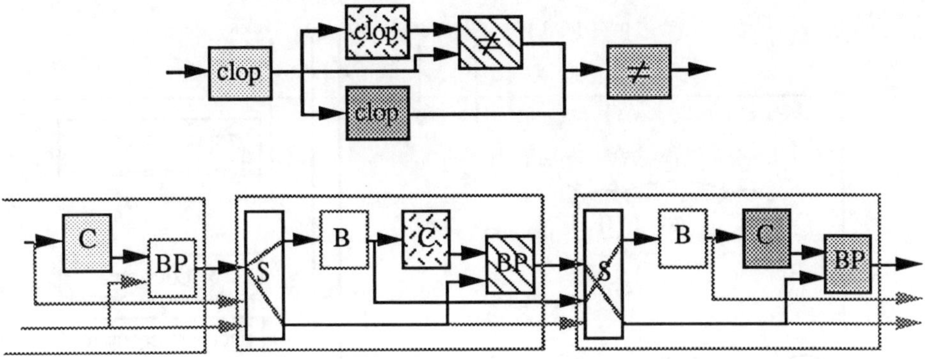

Figure 5.4.5 Forking, parallel operation and joining using CLPEs.

The block foreground_cnt in the datapath (figure 5.4.4) passes a 28 bit counter signal *fg_cnt* to the Master Module. The *result* signal is used to clock the counter with foreground pixels and the control signal from the Process Module control unit *ckr* (*clock_repeat_counter*) is used to reset the counter. Finally a control signal *dc* is passed to the control unit of the Process Module indicating if for the current pixel the image had changed due to the operation (*result* ≠ *original*).

The process module control unit.

The control unit of the Process Module is shown in figure 5.4.6.

It comprises a repeat counter to count the number of iterations. The counter is reset by the internal signal *clear_repeat_counter (clr)* and clocked by the internal signal *clock_repeat_counter (ckr)*. The full 12 bits can be read by the Master Module into its status register to return the number of iterations to the user (requirement i). The four least significant bits (*cycles*) are internally used in the control unit.

The x_counter and y_counter count the image size and are reset by the internal signals *reset_x (rsx)* and *reset_y (rsy)*. The pixel counter (x) is clocked by $\emptyset_1$ and the line counter (y) is clocked at the end of each line by the *reset_x (rsx)* signal. The 4 bit values *sizx* and *sizy* from the instruction register in the Master Module are decoded (see table 5.4.2) and their decoded values are used to compare with the counter values to give the end of line signal *(end_x)* and end of frame signal *(end_y)*.

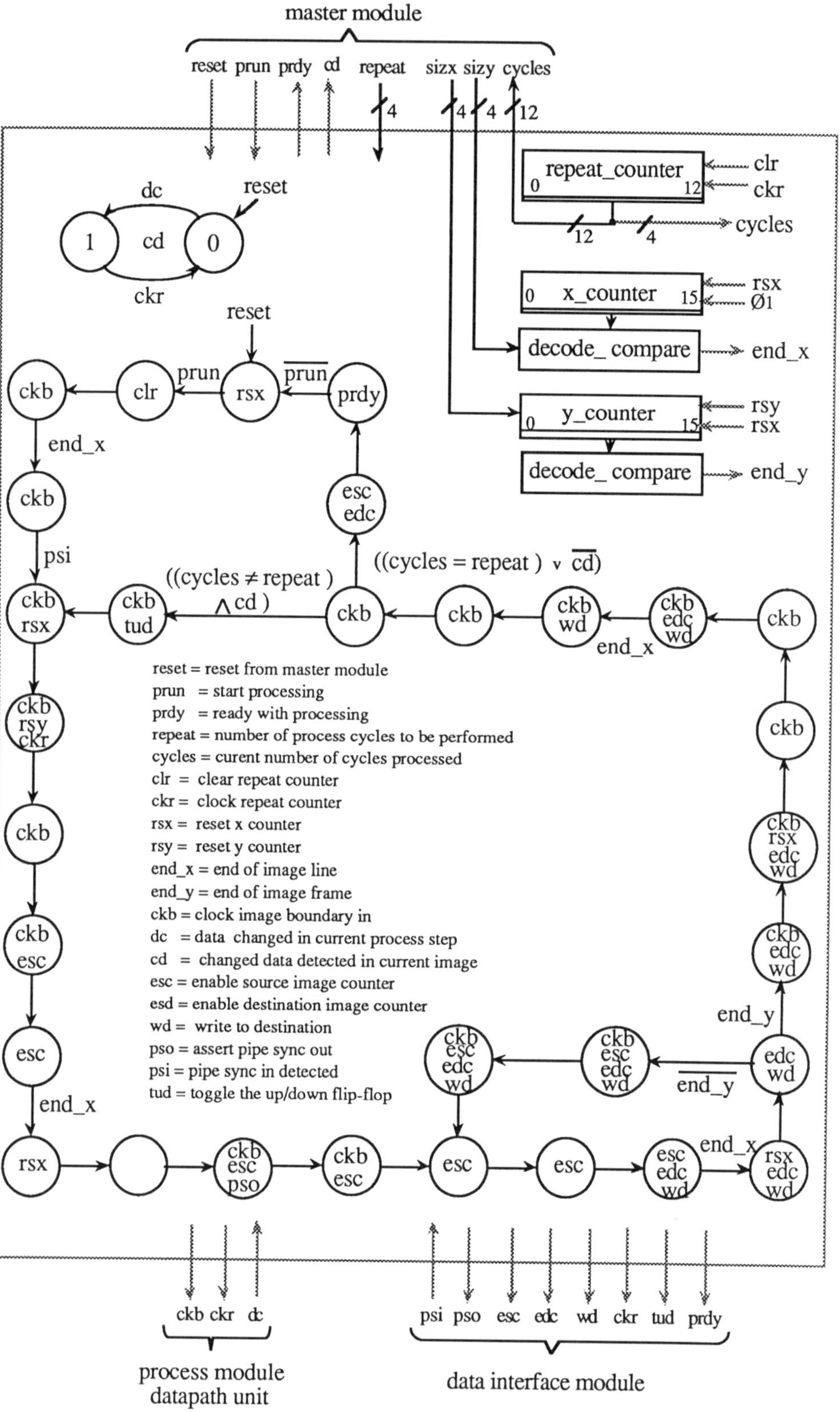

Figure 5.4.6 The process module control unit.

0000	32 bits	0101	1K bit
0001	64 bits	0110	2K bit
0010	128 bits	0111	4K bit
0011	256 bits	1001	8K bit
0100	512 bits	1010	16K bit

Table 5.4.2 Size settings of internal shift registers and external shift registers.

A 'change detected' flip-flop (*cd*) to detect idempotence (requirement g) is reset by the *reset* signal from the Master Module and by the internal signal *clock_repeat_counter (ckr)*. The flip-flop is set by the *detect_change (dc)* signal from the datapath of the Process Module.

The main statemachine of figure 5.4.6 is reset by the Master Module from any state to the *rsx* state by the signal *reset*. In this state the x_counter remains cleared. The statemachine is started by the *prun* signal of the Master Module. The main task of the statemachine is to insert boundary pixels in the Process Module datapath using the signal *ckb (clock boundary)* on the correct positions in the image. The exact timing of the various signals that the statemachine generates related to pixel positions in the image is shown in figure 5.4.7.

The statemachine is a synchronous statemachine, which means that each state transition takes one clock cycle ($Ø_1$, $Ø_2$). After the *run* command the statemachine clears the repeat counter *(clr)*, clocks in a line of boundary values*(ckb)* and waits for the pipe_sync_in signal *(psi)*. This signal from the Data Interface Module indicates that the data that enters from the Data Interface Module to the datapath input channels is valid. For a CLPE connected to an image memory, this signal should always be kept high. For a CLPE somewhere in a pipeline configuration this signal is connected to a *pipe_sync_out (pso)* signal of the previous CLPE in the pipeline.

After the *psi* signal arrived the statemachine steps through a number of states, resetting the y_counter, clocking the iteration counter and clocking the image source address counter(s) with the signal *(esc)*, This signal is passed to the Data Interface Module. the statemachine stays in the *esc* state until the x_counter indicates with *end_x* that the end of the image line is reached. The x_counter is reset and after stepping through a few states the statemachine signals with the pso signal the next CLPE in the pipe the validity of the pipe data. Consequently, after an asynchronous start-up of the pipeline, further processing is done clock-synchronously or systolic.

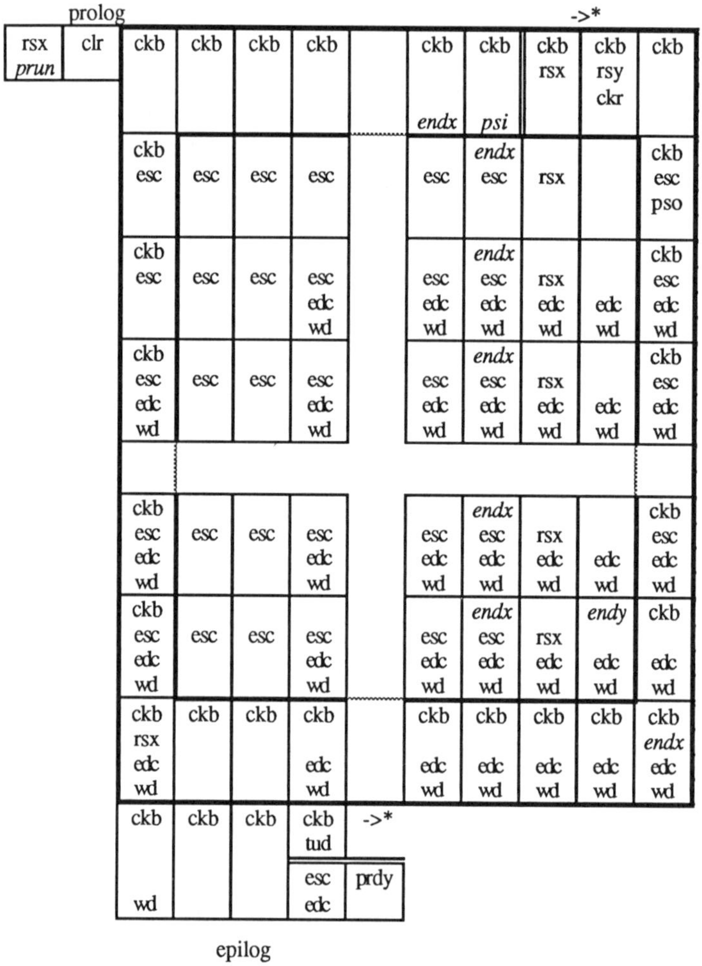

Figure 5.4.7 Control unit timing related to pixel positions in the image.

The loop in the lower right corner of the statemachine performs the handling of the non boundary lines of the image. After an image is finished, *(end_y)* followed by the last image line and *(end_x)*, it takes 5 more clock cycles to shut-off the pipeline and 7 clock cycles to start the processing of a new image. The bottom boundary pixels of the last image are reused as top boundary pixels for the new image.

The statemachine generates control signals for the Data Interface Module to manipulate the external bitplanes. This facility can be used for operating a single CLPE with an image memory as well as for the first CLPE and the last CLPE in the pipe to read and write up to 4 image memory bitplanes. Note that in each last state of an entire image iteration the signal *tud* (toggle_up_down) is given. This signal indicates the change in scan direction of the image memories. The signals for the image memory are: *enable_source_address_counter (esc)*, *enable_destination_address_counter (edc)* and *write_data (wd)*. The first two signals clock the address counters, the last signal clocks the destination data into the image memory. The signal *clock_repeat_counter (ckr)* can be used to reset the address counters.

The statemachine loops (performs iterations) as long as the datapath unit signals that the processing changes the image data (*cd* flip_flop) or that the number of repeats *(repeat)*, set in the instruction register of the Master Module, is unequal to the iteration repeat counter of the process control unit $((cycles \neq repeat) \wedge cd)$.

The data interface module.

Figure 5.4.8. shows datapath and control of the Data Interface Module. This module performs the interface function of CLPE to the other CLPEs in a pipeline configuration or to a maximum of four bitplanes of an image memory.

The module implements requirement a) "to function as a stand alone processor, able to address, read and write into image memory" and requirement m) "to iterate with upward as well as downward passes through the image to speed up recursive neighbourhood operations".

The instruction register of the Master Module provides the module with the bitplane address settings. These settings are:

Src1 , always one out of the bitplanes {a, b, c, d} and hence coded in 2 bits. *Src2*, *src3* and *msk* should be one out of the bitplanes {a, b, c, d} or are disabled and are hence coded in 3 bits. The destination *dst* can be written to any of the bitplanes {a, b, c, d} and is hence coded in 4 bits. The coding of *dst* in four bits facilitates the programmable storage of the result to more than one bitplane simultaneously. It can also be used for the programmable storage of the mask and the original image.

Requirement s) "a 40 pins case" made it impossible to include image memory address counters in the CLPE chip itself. The solution that is chosen was to output for each bitplane an address counter signal (ce_x) and a write enable signal (we_x). Moreover, a signal that can be used to reset the address counters *reset_C* and a signal that indicates whether the counters should count up or down *U / D* are provided.

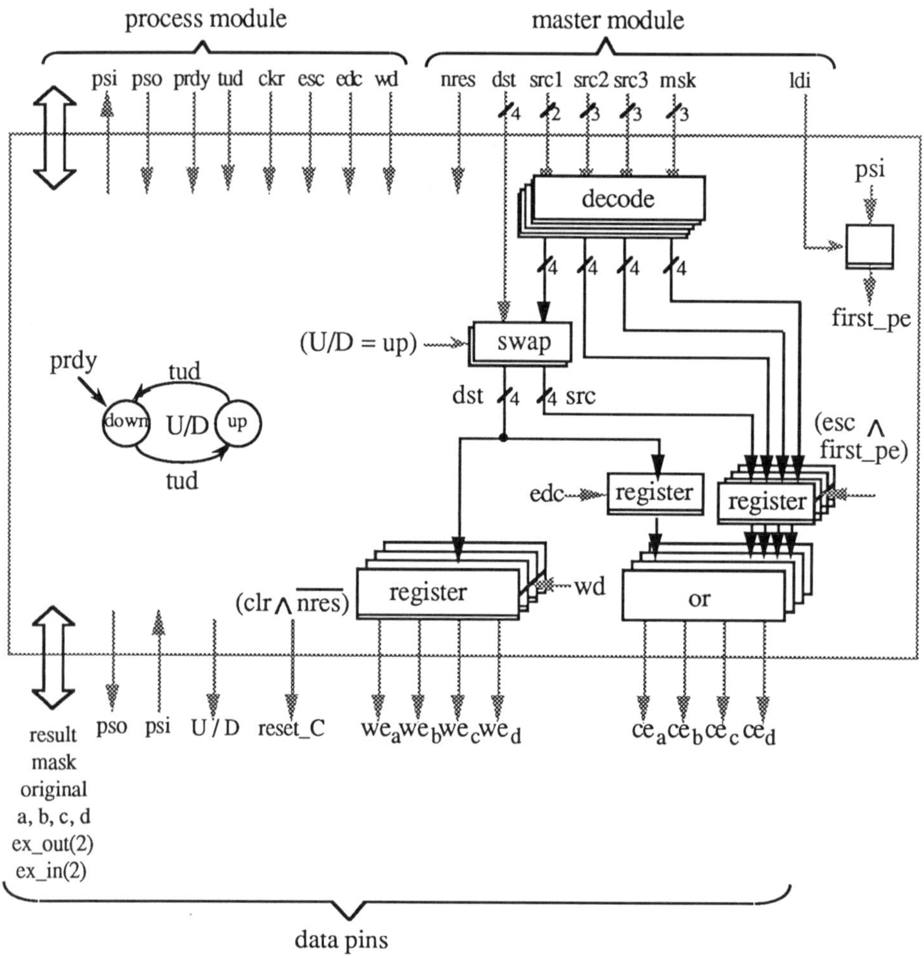

Figure 5.4.8 The Data Interface Module.

The master module provides the signals *first_PE* and *no_reset*.. The signal *first_PE* indicates that the CLPE is the first CLPE in the pipeline and that it hence should react on the enable source counter signal (*esc*) from the process module. The *no_reset*. signal can be used in combination with the iteration counter to perform the same operation on several images that are placed in consecutive order in the image memory; it prevents the reset of the external image address counters.

The U/D flip-flop is reset by the *process_module_ready (prdy)* signal from the process module and toggled by the *toggle_up_down (tud)* signal from the process module. The U/D signal from the flip-flop is used to switch the count direction of the external image address counters and to swap the source and destination bitplane definitions within the module. This allows continuously processing in both directions through the image. The actual timing of the address counter signals (ce_x) and a write enable signals (we_x) is performed by the control unit of the process module with the signals *esc*, *edc* and *wd*.

The following signals from the process module are only buffered by the bonding pad buffers in the module; the output signals: *result*, *mask* and *original*, the input signals *a*, *b*, *c* and *d* , the external shift register signals *ex_out* (2) and *ex_in* (4) and the pipe sync signals *ps_in* and *ps_out*.. Note that the various clocked and unclocked bonding pad buffers are not drawn in figure 5.4.8. Figures 5.4.9 shows the connection and timing of the possible external shift registers for image sizes larger than 512 in the X direction. Image sizes larger than 512 in the y direction are handled in the CLPE by the y_counter.

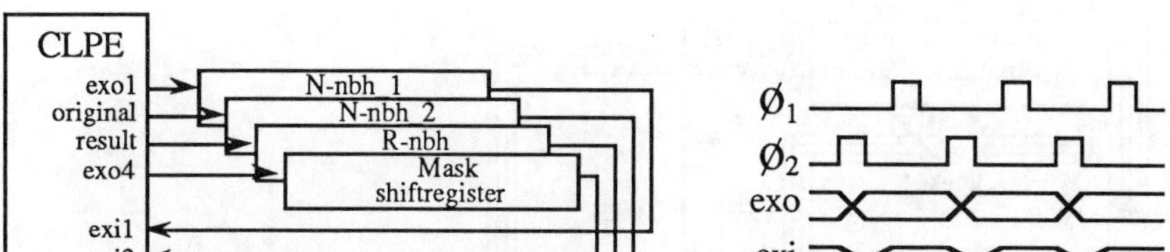

Figure 5.4.9 External shift register connection and associated timing.

Figures 5.4.10 shows a number of CLPEs in a pipeline configuration with first and last CLPE connected to image memory. The timing of the CLPE inputs (a,b,c,d) and its outputs (result, original, mask) is equal to the timing of the external shift registers (figure 5.4.9), with CEx occurring on clock pulse $Ø_1$ and WEx occurring on $Ø_2$.

The first CLPE and the image memory module can also be used as a stand alone unit: The CLPE is capable of addressing any bitplane on any start position in the memory and independent for reading and for writing. The relative independent addressing capability of the CLPE allows a variety of memory configurations. Gradually more CLPEs can be attached to the pipeline. Generally, the pipeline should be longer than the longest sequence of cellular logic operations, including unfolded iterations of iterative cellular logic operations. If the pipeline is shorter, frame re-circulation can be applied through the image memory. In that case, the image can be processed in upward and downward scans. Figure 5.4.10 shows a feedback from the result of the first CLPE to the image memory. This can be used for short operations such as fast copying of images through the image memory. The capability of each CLPE to address four output bitplanes for writing allows two more intermediate result taps to be placed on the pipeline. This may speed up operations that are shorter than the pipeline length.

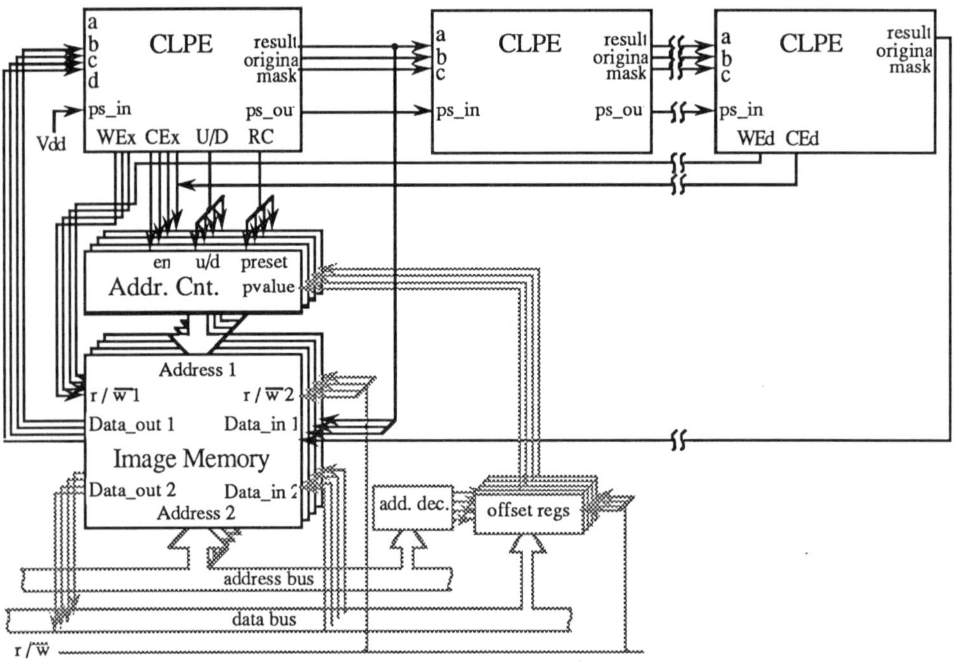

Figure 5.4.10 CLPEs with memory and pipeline connections.

The instruction interface module.

Figure 5.4.11 shows the Instruction Interface Module. Restriction s) "*a 40 pins DIL case*" made it necessary to download WLA words and instructions to the CLPE in a serial way. CLPE status words on requests by the host are also transferred sequentially. To limit the number of connections, the internal shift registers for instruction / status and WLA_words were directly placed in the Master Module and

Mask Module respectively. As a consequence the independence of host access and processing is the concern of the Mask Module and Master Module.

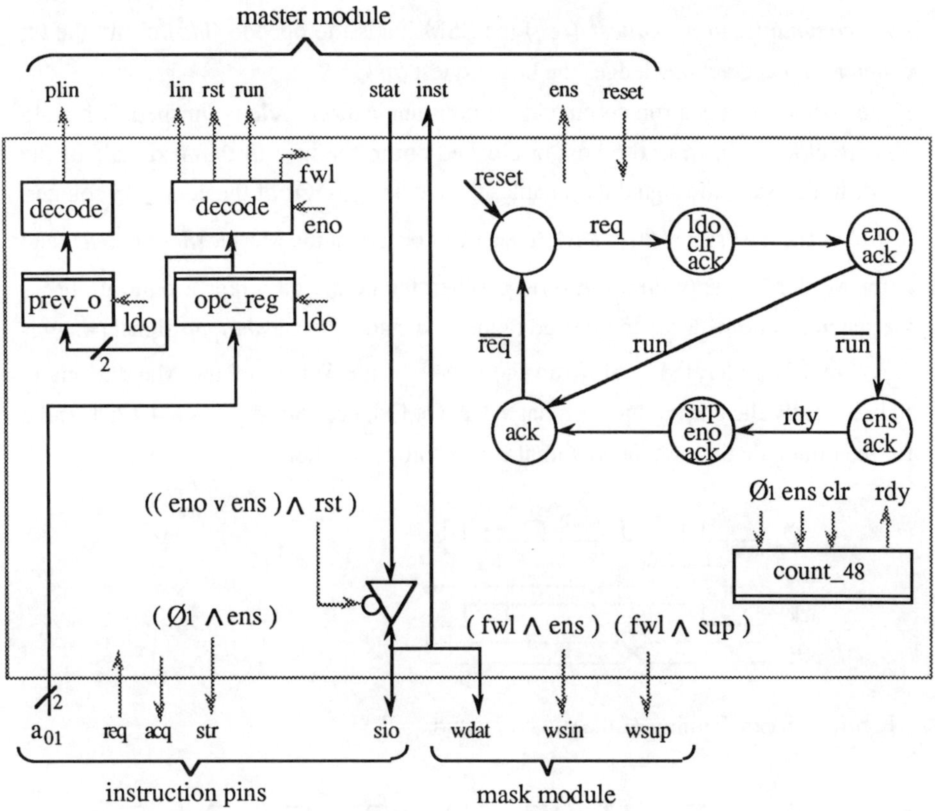

Figure 5.4.11 The Instruction Interface Module

The type of host I/O cycle is given by the two control pins a_{01} that are clocked into an opcode register on the signal *load_opcode (ldo)*. The two bits are decoded to four signals indicating the I/O mode; *load_instruction (lin)*, *read_status (rst)*, *fill_wla (fwl)* and *process_the_last_instruction (run)*. The signals are enabled to pass to the Master Module and Mask Module on the signal *enable_opcode (eno)*. As the Master Module needs to know when a previous opcode was a load instruction the signal *previous_opcode=load_instruction (plin)* is made.

The connection of the *serial_I/O* line *(sio)* to the *status_output* line *(stat)* is enabled by a read status *(rst)* by the host and the controlling statemachine gives the signals

enable_opcode (eno) or *enable_shift (ens)*. In all other cases, the *sio* line is connected to the *instruction_input* line *(inst)* and the *wla_data_input* line *(wdat)* .
The synchronous statemachine (SSM) is reset by the Master Module. On a *host_communication_request (req)* the SSM loads the opcode *(ldo)* clears the bit counter *(clr)* and acknowledges the host request *(ack)*.
If the host request is a run command the communication cycle is finished. If not, 48 bits are clocked in from the host or clocked out to the host in the next state of the SSM. In this state the signal *ens* enables the clocking on $\emptyset_1$ of the count_48 counter, it strobes the host $(str = \emptyset_1 \wedge ens)$, the shift registers in the Master Module *(ens)* and in the Mask Module $(wsin = ens \wedge fwl)$. When the count_48 wraps around it allows the statemachine with *rdy* to proceed to the next state where a *shift_up* signal $(wsup = sup \wedge fwl)$ is generated to shift up the masks in the WLAs of the Mask Module. Subsequently the communication handshake is finished. See figures 5.4.12a,b and c for the timing that is associated with the host communication.

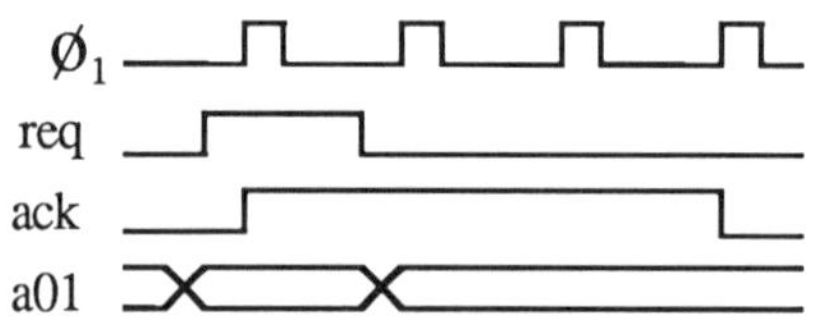

Figure 5.4.12a Timing of the run command.

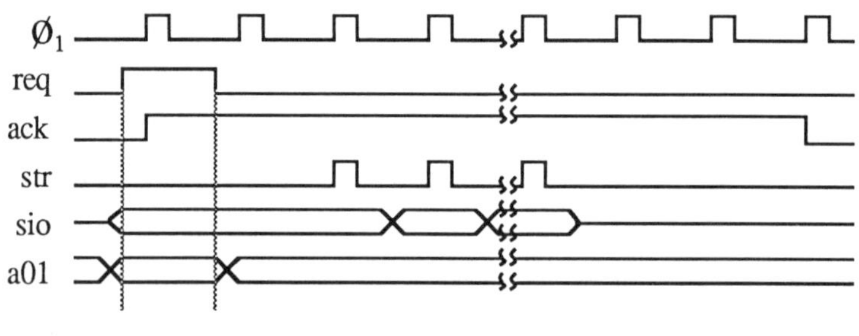

Figure 5.4.12b Timing of an instruction / mask write command.

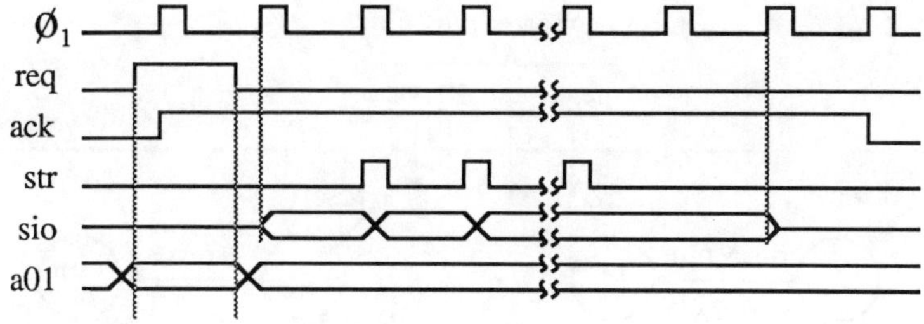

Figure 5.4.12c Timing of a status read command.

The master module.

Figure 5.4.13 shows the implementation of the Master Module. The main datapath is the shift register - instruction register pair. Sequential input to the shift register is the instruction signal *(inst)* from the Instruction Interface Module. Sequential output from the shift register is the status signal *(stat)* to the Instruction Interface Module. A shift is performed on the *enable_shift* signal in both the *load_instruction* mode and *read_status* mode $((ens \wedge lin) \vee (ens \wedge rst))$.

The CLPE status that is made available to the host is the number of iteration cycles the CLPE has performed since the last run command *(cycles)*, the number of foreground pixels in the result pixel flow since the last run command *(fg_cnt)*, a bit indicating if the image had changed in the last iteration *(cd)*, a bit indicating if the processor is CLPE running *(prun)*, and a bit indicating if the CLPE has been enabled to generate an interrupt to the host when it is finished *(ine)*. The CLPE status is parallel loaded into the shift register on the status read signal *(rst)* from the instruction Interface Module.

The instruction register is loaded from the shift register on the instruction load *(ldi)* signal from the main statemachine of the module. The instruction signals are the image size *(sizx, sizy)* the settings of the Boolean function generators *(bfg1, bfg2, bfg3)* the number of iterations to be performed *(repeat)*, the operation code for the Mask Module *(opc)*, the selection of source and destination channels of the CLPE *(src1, src2, src3, msk, dst)*, the image boundary value *(ib)*, the control signal to disable the resetting of the address counters *(nres)* and finally the interrupt enable signal *(ine)*. The master pins *(reset, test,* $\emptyset_1, \emptyset_2$) are buffered in the Master Module, used internally and passed to all other modules.

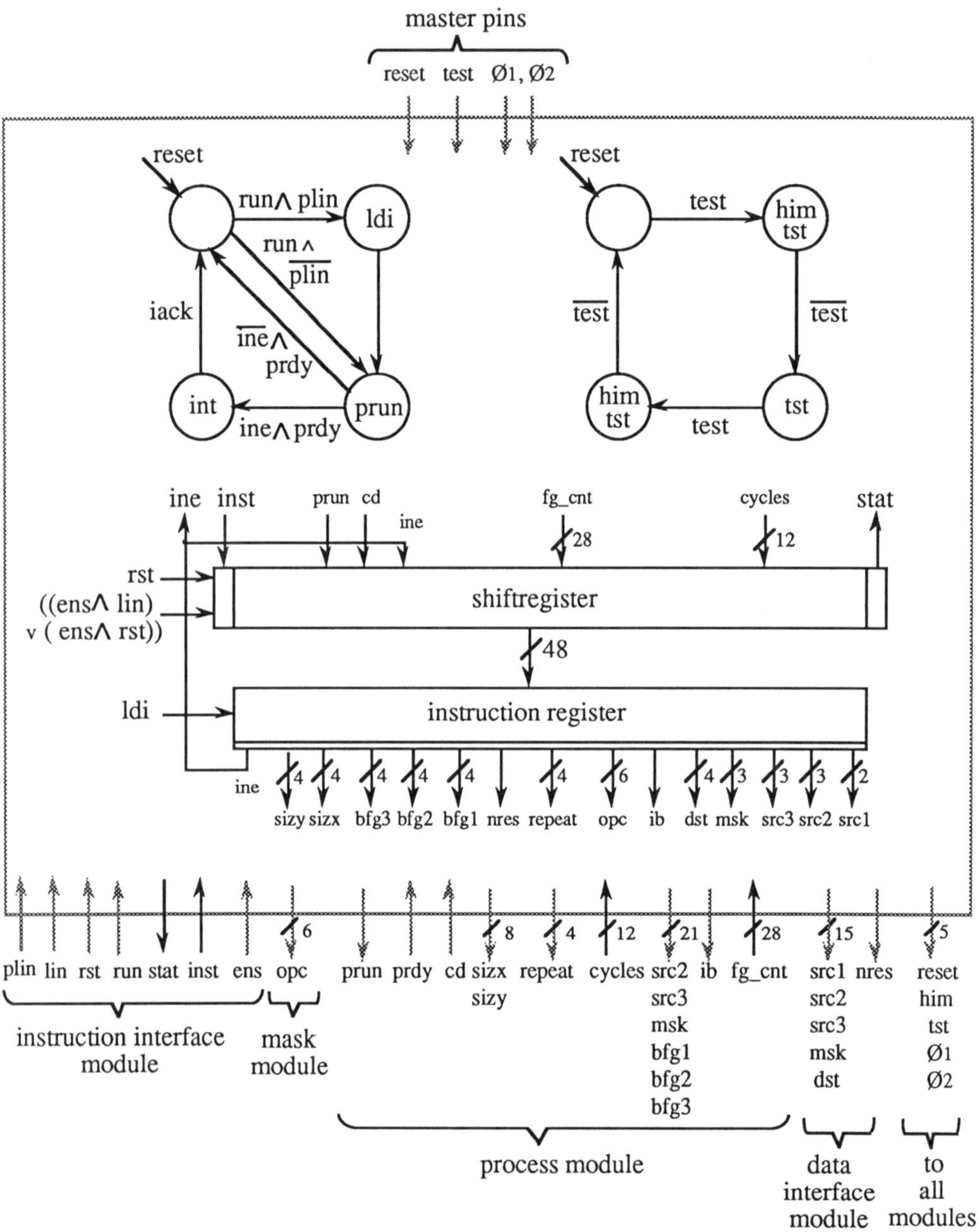

Figure 5.4.13 The Master Module.

The CLPE is capable to interrupt the host provided that it is enabled by the host through the setting of *ine*. The main statemachine shows this, it implements requirement n) "*to be started and reset by a host and to interrupt this host when processing is finished*". A *reset* brings the main statemachine from any state into the initial state. If the command before the *run* command was a load instruction command, signalled by the *previous=load_instruction (plin)* signal from the instruction Interface Module, the instruction is loaded from the shift register into the instruction register with the *load_instruction (ldi)* signal. Directly after this, in the next state, the Process Module is started with the *prun* signal. If the previous command was not a load instruction, no new instruction is loaded and the old instruction is repeated. When the Process Module indicates that it is ready *(prdy)*, an interrupt is given to the host *(int)*, if the interrupt was enabled *(ine)*. If the interrupt was not enabled the statemachine proceeds directly to the idle state. When an interrupt was given the statemachine waits for the interrupt acknowledge signal *(iack)* before it returns to the idle state.

5.4.5 Testing facilities.

Requirement r) "*the correct functional behavior of the VLSI circuit should be made testable through its bonding pads (pins)*" together with restriction s) "*a 40 pins case*" made it necessary to reserve one pin on the IC to switch the CLPE into test mode. The test statemachine in the Master Module shows that on the assertion of the test pin a state is reached in which first all non-master pins of the CLPE are placed in the high impedance mode by sending a *him* signal to all modules. At the same time, this freezes all internal states of the CLPE. When the test pin is released again, the high impedance state of the non-master pins is cancelled and the CLPE is in test mode. Once again asserting the test pin brings the CLPE in normal mode back again via an intermediate high impedance state. These high impedance states can be used to switch signals on a testing device. Even manual switching is possible.

The internal testing system of the CLPE is based on the scan path concept; the Level Sensitive Scan Design (LSSD) (Eichelberger and Williams 1978). In the LSSD method all latches and registers are replaced by parallel loadable shift register cells. In normal mode, these devices operate in parallel load mode, in test mode the contents of the registers can be shifted out and new data can be shifted in. A number of registers are chained to a scan path with begin and end of the chain connected to an IC pin. In this way, the internal state of the CLPE can be examined by shifting out all register values and, moreover, be preset with new values to bring the CLPE in a totally different state if necessary. However, in some complex combinatorial situations there are no registers near the points that are nevertheless interesting test points. For this purpose, the LSSD method was extended with some special solutions based on the line scan cell (Schot 1988). See figure 5.4.14 for the normal scan cells in a scan line and figure 5.4.15 for the line scan cell.

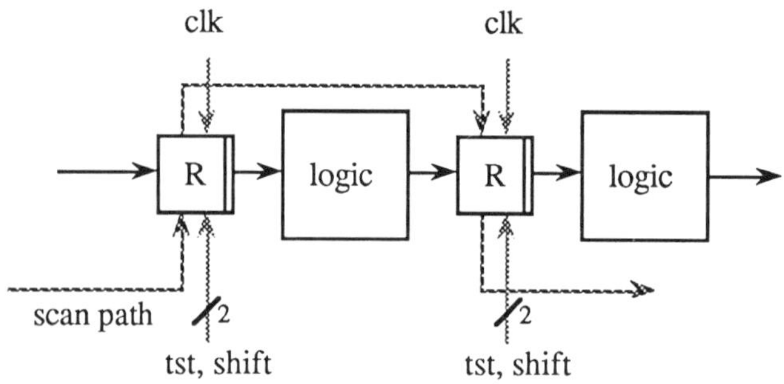

Figure 5.4.14 The LSSD method.

In a line scan cell, in parallel to a signal line a scan register cell is placed with its input connected to the signal line and its output connected to a selector. In normal operation the value of the line is clocked in at each clock pulse. In the test mode the line value at the last clock pulse can be examined by shifting it out through the scan path. In the test mode a new line value can be inserted into the circuit using the selector to select the register value instead of the signal line value.

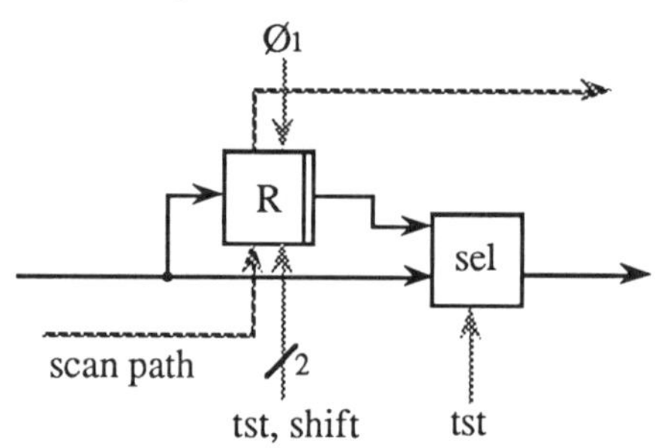

Figure 5.4.15 The line scan cell.

The line scan cell is used for the internal testing of the PLAs that are used to realize the synchronous statemachines of the control and the counters. See figure 5.4.16.

Special solutions are made for the dynamic delay cells of the four line shift registers of the CLPE and the WLA. When shifting in test mode all four shift registers are connected in one scan path. To hold the data in test mode, chunks of 64 bits are circular connected, clocking the data around.

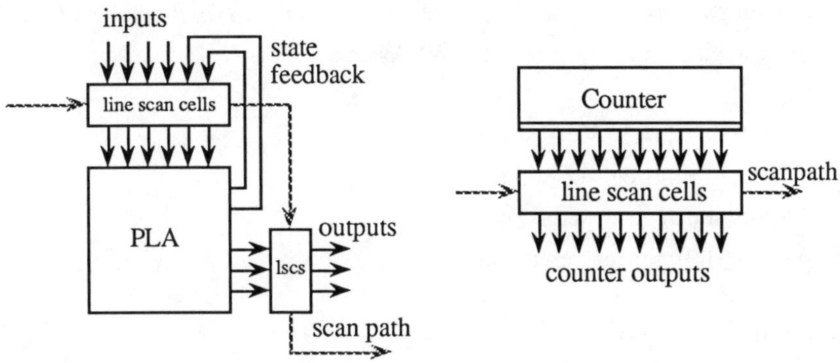

Figure 5.4.16 The use of line scan cells for PLA and counter testing.

To test the WLA a dynamic shift register is placed on top of the WLA. The WLA terms that are shifted in from the bottom of the WLA and shifted out on top are now picked up by the test shift register and shifted out into a scan path.

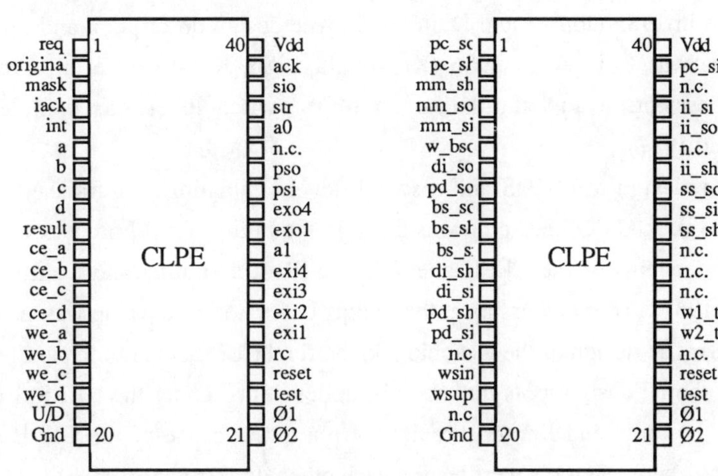

Figure 5.4.17 CLPE pin lay-out in normal mode and in test mode.

Figure 5.4.17 shows the CLPE pin lay-out in normal mode and in test mode. Each scan path has a serial in pin *(xx_si)* a serial out pin *(xx_so)* and a scan path shift pin *(xx_sh)*. The scan paths are: the scan path of the master module *(mm)*, the scan path of the process module control unit *(pc)*, the scan path of the process module data path *(pd)*, the data_interface scan path *(di)*, the instruction interface scan path *(ii)*, the line shifters scan path *(ls)*, the image boundary logic scan path *(bs)* and the WLA scan path *(w)*. The pins for the testing of the WLAs are:

WLA_shift_in *(wsin)*, WLA_shift_up *(wsup)*,
WLA_bottom_shiftregister_serial_out *(w_bso)*,
WLA1_top_shiftregister serial _out *(w1_tso)*
WLA2_top_shiftregister_serial_out *(w2_tso)*.

5.4.6 Design history.

The roots of the CLPE design laid in the logic table module of the Delft Image Processor DIP1 (Gerritsen and Aardema1981). A stand alone VME bus implementation of this module was realized in 1985 (Boekamp et al. 1985). The CLPE design was influenced by the research on the performance differences between processor arrays and pipelines; a similarity between the CLIP4 PE (Duff 1982) design and the CLPE design was pursued.

The first design of a cellular logic PE based on sets of 3 x 3 structuring elements was performed in 1985 (Jonker and Duin 1985). The design description and functional simulation was finished in 1986 (Kraaijveld 1986, Kraaijveld et al. 1986). An NMOS description and simulation on MOS switch level was completed by (Kraaijveld 1987).

A circuit redesign to CMOS and a switch level simulation was finished in 1988 (Schot 1988). A CMOS test chip was made in the Philips C5TH process with some major components of the CLPE; the WLA, a PLA, a counter and a shift register (Venema 1987). After processing at the Philips Laboratories, the chip was tested (van Dijk 1988). The design of the test chip and the final CLPE chip ran in parallel to the design of the IC design tools that have been developed under the NELSIS project. Many of the tools and library circuits were a direct cause of the CLPE design. Moreover, design and tools were testing each other (Jonker et al. 1988b).

A separate WLA chip (see also section 5.1) was designed in 1989 (Schmidt et al. 1988), processed in the ES2 process and tested in 1991 (De Zwart 1991).

A functional simulator, embedded in an Image Processing Package (TCL-Image 1990), for 2D and 3D cellular logic operations based on sets of structuring elements was finished in 1989 (Klop 1989).
The final CLPE chip was processed in 1990 at the Philips Laboratories in Hamburg. The chips have been tested in 1991 on a VLSI test system (van der Knijff 1991) and a first implementation was made using a Real-time Pipelined Image Processor (Imaging-Technology 1990).

5.5 A real-time pipeline for low level image processing.

5.5.1 Initial Grey Value Slice design.

The possibility to upgrade the CLPE in a later stage for bit wise grey value processing was one of the requirements in the original design of the CLPE. For this purpose a fourth input was created. A simplified block diagram of the CLPE is shown in figure 5.5.1.

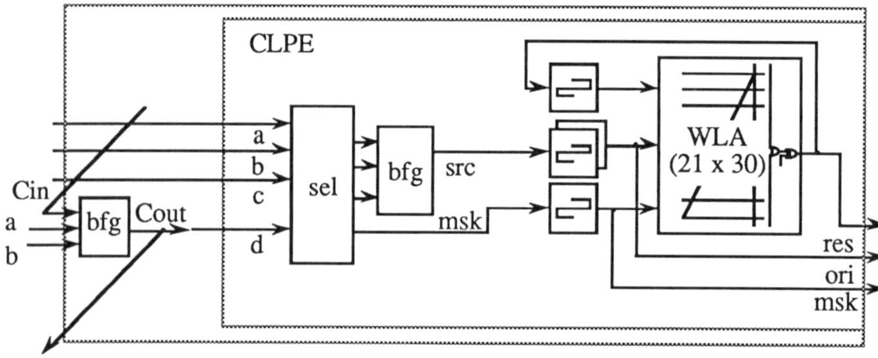

Figure 5.5.1 A simplified block diagram of the CLPE.

In this diagram the output Boolean function generator is replaced by a programmable inverter. With the *msk* image fed through the WLA this inverter was found to be sufficient to perform any Boolean post processing between *src* and *msk* image. Moreover, the Majority Vote Unit is not drawn here to simplify the diagram.

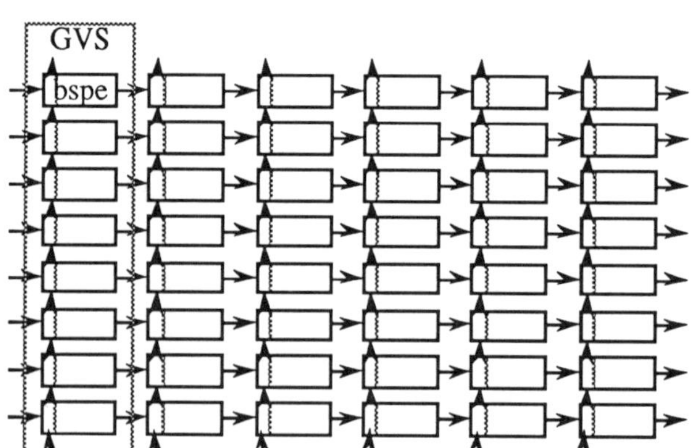

Figure 5.5.2 A pipeline for grey value image processing.

Processing grey value images of *b* bits could be performed with a stack of *b* CLPEs and appropriate carry-over circuits (Duin and Jonker 1988). For this purpose *b* Boolean function generators and a ripple carry circuit producing a carry vector could be connected to the fourth input. The stack of extra Boolean function generators could be used as first half adder, the input Boolean function generators of the CLPEs could be used as second half adder, in case of addition of two images. A stack of *b* extended CLPEs is called a grey value slice (GVS).

A pipeline of *l* grey value slices can be used to process *l* grey value instructions or iterations of grey value operations. (See figure 5.5.2) The low level grey value operations that this pipeline preferably has to support can be classified in the following way (Adapted from Komen and Duin 1989):

Low Level Operation types:	Single Bit	Word: Bit wise	Word: Bit neighbhood	Word: Bit Recursive Neighbourhood	Word: Global
Monadic Point	Not	Abs	Bit Shift	Incr, Decr, Sqrt	Sine, Ln,..
Dyadic Point	And, Or, Xor,...			Add, Sub, Threshold Mul-, Div-, Mod-pass	Max, Min
Monadic Neighbourhood Morphological	Ero, Dila, Contour,... Image Shift.	Thr decomposed filtering, Image shift.		Gradient filters	
Monadic Neighbourhood Statistical	Binary Rank (= Majority vote)	Threshold decomposed rank filter		General Convolution Gauss	LMax, LMin, Rank Filtering Max_Min ops.
Monadic Recursive Neighbourhood Morphological	Skeleton Pass			Sobel, Laplace	
Monadic Recursive Neighbourhood Statistical	(Recursive Binary Rank Filter pass.)			Fast Uniform, Laplace	Recursive Rank Filter pass.
Dyadic Point Neighbourhood					
Dyadic Point -Recursive Neighb. Morphological	Propagation pass, Anchored skeleton pass	Grey value propagation pass			

Dyadic Point -Recursive Neighb. Statistical					Distance Transform pass
Monadic Image Global	Global Or			Pixel Count	FFT, Affine transform
Dyadic Point -Image Global				Histogram, Eql, Cst	

Table 5.5.1 Classification of low level grey value operations.

For comparison purposes, the binary and cellular logic operations as performed by the CLPE are included in the table. A differentiation is made between morphological and statistical operations. For morphological operations, the base of the operation is form detection within a neighbourhood, for statistical operations the base of the operation is event counting within a neighbourhood. These two classes are represented by the WLA and the MVU of the CLPE for the binary case.
(Komen and Duin 1989) used the CLPE extended with carry logic as shown in figure 5.5.1, in a performance comparison with grey value processing on the CLIP4. In this comparison, the assumption was made that the ripple carry circuit is able to ripple the carry from LSB to MSB or vice versa in one clock cycle. The carry propagation plays a role in all bit neighbourhood and bit recursive operations. The spatial recursion of the carry propagation is similar to the two dimensional recursion in the cellular logic propagation operation. The most evident operations using carry propagation are the:

The addition:

$$\{\ \forall_{i=0}^{b} \mid g[i] \leftarrow (a[i] \wedge b[i])\ ;$$
$$t\,[i] \leftarrow (a[i] \neq b[i]);$$
$$c[i+1] \leftarrow (g[i] \vee (\ t[i] \wedge c[i]\ ));$$
$$r[i] \leftarrow (t[i] \neq c[i])$$
$$\}$$

with
g: the carry generation condition,
t: the carry transmission condition and
r: the result of the operation.

(5.5.1)

The subtraction:

$$\{\ \forall_{i=0}^{b} \mid g[i] \leftarrow (a[i] \wedge \overline{b[i]})\ ;\\ t\,[i] \leftarrow (a[i] = b[i]);\\ c[0] \leftarrow 1;\quad c[i+1] \leftarrow (g[i] \vee (\,t\,[i] \wedge c[i]\,));\\ r[i] \leftarrow (\,t[i] \neq c[i]\,)\\ \}$$

(5.5.2)

Thresholding:

$$\{\ \forall_{i=0}^{b} \mid g[i] \leftarrow (\,a[i] \wedge \overline{b[i]}\,)\ ;\\ t\,[i] \leftarrow (\,a[i] = b[i]\,);\\ c[0] \leftarrow 1;\quad c[i+1] \leftarrow (\,g \vee (\,t \wedge c[i]\,));\\ r \leftarrow c[b+1]\,)\\ \}$$

(5.5.3)

(The threshold $r \leftarrow (a > b)$ is the carry out propagation of the subtraction operation.)

Incrementing: $\{\forall_{i=0}^{b} \mid c[0] \leftarrow 1;\ c[i+1] \leftarrow (a[i] \wedge c[i]\,);\ r[i] \leftarrow (\,a[i] \neq c[i]\,)\ \}$

(5.5.4)

(Incrementing is adding the carry to a.)

The bit shift is the non recursive bit neighbourhood operation:

$$\{\forall_{i=0}^{b} \mid a[i+1] \leftarrow a[i]\ \}$$

(5.5.5)

Many more complex operations and filters are based on these basic operations. Multiplication, division and integer square root can all be considered as a fixed number (*b*) of iterations of a basic pass. For example a basic multiplication pass is executed using a double word length result variable and comprises a conditional addition and a shift of the temporal result:

Multiplication:

$$\{\forall_{i=0}^{b} \mid r_l \leftarrow a;\ (\,(r_l[0] = 1) \mid r_h \leftarrow \mathrm{add}(r_h, b));\ \mathrm{shiftright}\ (r_l r_h)\ \}$$

(5.5.6)

To enable a variety of carry-over conditions a Boolean Function Generator with four inputs (a, b, Carry_in_Low, Carry_in_High) was used to make the carry function fully programmable.

5.5.2 Enhanced Design of a Grey Value Slice.

A prominent observation in the comparison (Komen and Duin 1989) was that iterative operations based on shifts and additions take too many grey value slices, because both for a bit shift (using the carry circuit) and for the addition a grey value slice was needed. Moreover, *b* iterations are needed to perform one b bit multiplication. A Look-Up Table, however, is capable to perform all point wise grey value functions in one clock cycle. Figure 5.5.3 shows a design of a Grey Value Slice based on a Look-Up-Table with slightly modified CLPEs.

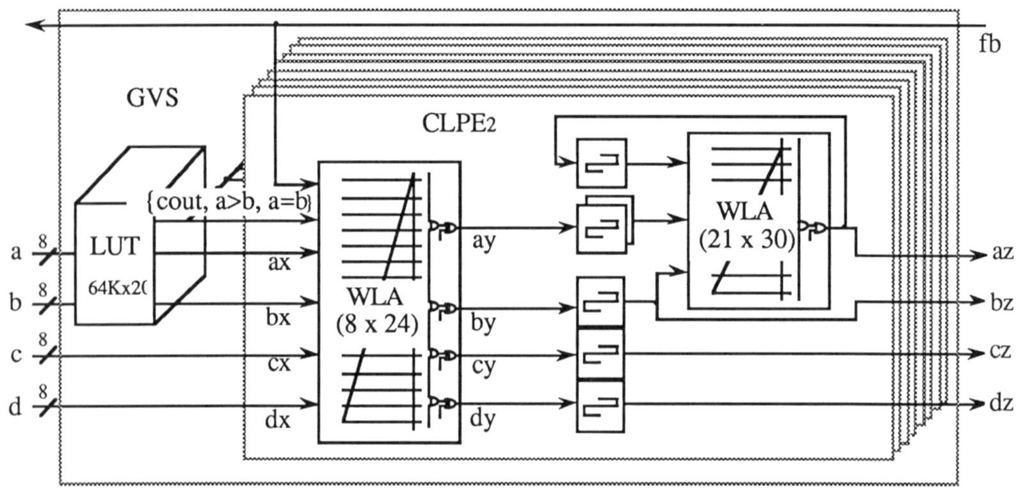

Figure 5.5.3 A Greyvalue Slice.

The modification of the CLPE is twofold: Firstly, as variables in a pipeline system are realized by the independent data channels of the pipeline, two more variables are introduced to relax the limitations of the original CLPE. Secondly, the input circuit is changed in such a way that each data channel can be a Boolean combination of any eight CLPE inputs. Four inputs being the normal data channels and four being additional global control signals (equal for all CLPEs in the slice), to allow local data dependent control. Secondly, the fixed feed through of the *original* channel is omitted. Using the input WLA of the CLPE the *original* can always be fed through using one of the new auxiliary data channels.

Basic arithmetic operations, using the LUT, such as additions, multiplications, bit shifts, maximum and minimum operations, can be performed in one clock cycle. Multiplication may yield a 16 bit result.

General convolution, using the neighbourhood WLA to select neighbourhood points, can be performed in eight iterations, i.e. eight Grey Value Slices. Any operation based on general convolution in a 3 x 3 neighbourhood can hence be realized.

Local Maximum and Local Minimum functions in a 3 x 3 neighbourhood can also be obtained by eight comparisons performed by eight successive LUT operations (in general by k^2-1 comparisons for a k x k neighbourhood). Using the LUT a compare and exchange function can be performed. A serial sort in a k x k window can be used as a straight forward rank filter algorithm. See figure 5.5.4. However, in the worst case $k^2(k^2-1)/2$ iterations are needed to find an arbitrary rank. Moreover, the k x k partial orderings have to be stored in temporal variables. Yielding 36 iterations and 9 temporal variables for a 3 x 3 environment.

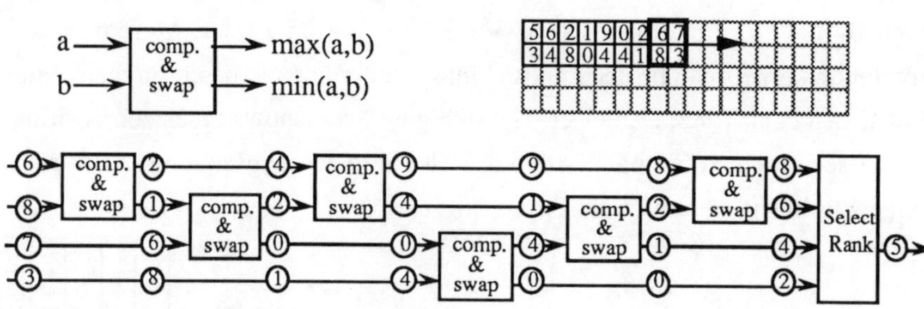

Figure 5.5.4 Rank filtering by sorting.

Fortunately, many k x k convolution based filters can be split in a convolution of k x 1 with 1 x k (Groen et al. 1988). Using this row and column separation approach the uniform filter, gaussian filter, the local maximum and minimum filters can generally be performed in 2(k-1) stages.

Rank filtering in a 3 x 3 environment can be programmed with a set of local maximum, minimum and median operations, first in horizontal direction, then in vertical direction (Jonker et al. 1989):

Rank[i] <- Vertical Rank[i div 3] (Horizontal Rank[i mod 3]). (5.5.7)

Figure 5.5.5 shows the Local Maximum and Local Median filter in 5 pipeline stages.

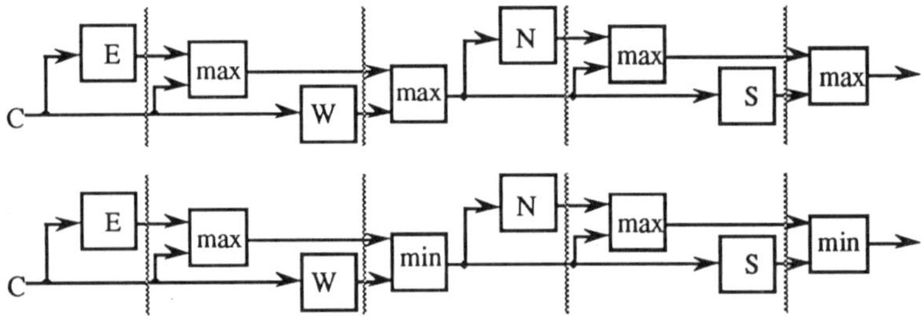

Figure 5.5.5 Local Maximum and Local Median Filtering using row and column separation.

Large kernel filtering for operations such as shading compensation using Local minimum and Local Maximum operations using a k x k kernel can generally be solved by (k+1)/2 iterations with a 3 x 3 convolution kernel. Moreover, many convolution kernels can be decomposed into smaller kernels based on the recursive form (Groen et al. 1988, Danielson 1990). Figure 5.5.6 shows the decomposition of a 5 x 5 separated Sobel pair (Åström 1990) into ten subsequent additions and subtractions in ten grey value slices.

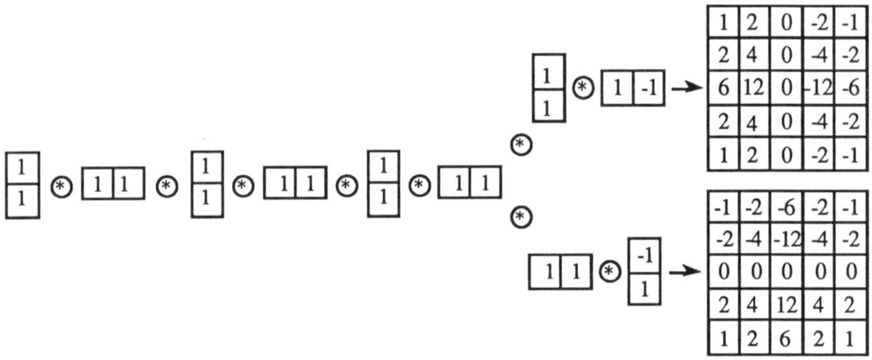

Figure 5.5.6 Decomposition of a 5 x 5 Sobel kernel pair.

Note that for all these filters the inherent neighbourhood parallelism of the normal neighbourhood and the recursive neighbourhood of the CLPE is unusable for grey value operations, since all word wide operations can only be performed as dyadic point operations performed by the LUT. The direction selection in the neighbourhood, however, is performed by the neighbourhood WLA.

Some more complex algorithms will be discussed now.

Histogram collection is quite cumbersome in a Pipeline, since the pipeline structure does not allow a variety of data structures. However, the counting facility of the CLPE on output channel *az* can be used for this purpose. Note, that this requires for e.g. a 128 value histogram 128 CLPEs or frame recirculation for smaller Pipelines. Each LUT is used to select separate grey values values which are sent through the WLAs to the output channel *az*, where the number of foreground bits are counted and the image is further sent to dust. The original image is passed through channel *b* and tapped by subsequent slices for further detection. Histogram based thresholding methods can be further processed in the PipeLine host, after which the threshold value is sent to the PipeLine. A 128 value histogram takes 16 GVSs. In addition, the interrogation of the CLPEs costs at least (48 x 8) clock pulses to upload the status of the 16 slices. This will increase to a few thousand clock cycles if the communication overhead is taken into account.

The **distance transform** of a binary image can be found with the (Borgefors 1986) algorithm that produces the distance transform by means of a 2-pass recursive filter operation. See figure 5.5.7. The filter size 3 x 2 or 5 x 3 depends upon the desired accuracy, the result is an image where each pixel contains the minimum 'chamfer' distance to a background pixel. The Borgefors distance is indicated with BDT(d_1,d_2,d_3), the grid constant is d_1, the diagonal distance is d_2 and a knights move is d_3. A local correct euclidean distance is then BDT(1,$\sqrt{2}$,$\sqrt{5}$), with as a good rational approximation BDT(5,7,11), yielding a hexadecagonal isodistance shape.

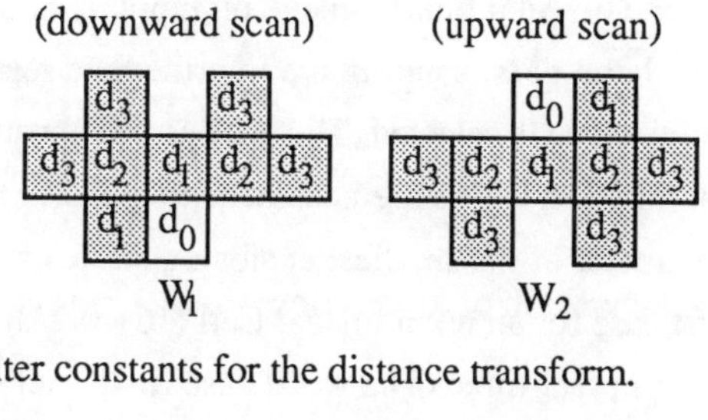

Figure 5.5.7 Filter constants for the distance transform.

The algorithm for the 3 x 2 distance transform BDT(5,7,∞) yields an octagonal isodistance shape and can be implemented on the PipeLine using the recursive neighbourhood of the CLPEs. The downward pass:

$$C_r = \min (C_n, W_r+5, NW_r+7, N_r+5, NE_r+7) \tag{5.5.8a}$$

is applied in the form:

$$C_r = \min (NE_r+7, \min (N_r+5, \min (NW_r+7, \min (C_n, W_r+5)))) \tag{5.5.8b}$$

with: C_o, Resulting Central pixel, C_n, Central pixel in the normal neighbourhood, W_r , West pixel in recursive neighbourhood, etc.

and yields an intermediate result after eight slices.
The upward pass with window W_2 should be performed from intermediate image memory using the CLPEs to address this memory in opposite direction. The distance transform takes 16 GVSs and one frame time extra due to the intermediate memory.

In the **constraint distance transform**, a binary constraint or obstacle image is used to disable the change of distance values on the pixel positions of the obstacles. The constraint image can be input through channel *by* in the neighbourhood WLA to generate the value ∞ (infinite) on constraint positions without further delay.

In the **pseudo euclidian skeleton** (Verwer 1988) the distance transform of the image is used to determine the order in which the pixels are conditionally eroded, using the breakpixel conditions as defined in section 4.1.
When the distance image is set on input *b* (see figure 5.5.3), the LUT is programmed to select all pixels with the lowest distance, yielding a binary image on *by*. The input WLA ANDs this with the original binary image on input *ay* and the Neighbourhood WLA performs one skeleton pass, sending the intermediate result to the next GVS, where the next lowest distance is selected. This process is repeated until all distances are processed. In general 16+d GVSs are necessary to perform the pseudo euclidian skeleton, with d the diameter of the smallest enclosing circle of the thickest object in the image. In the worst case for an n x n image: Ceil $(16+n\sqrt{2}/2)$ GVSs are necessary to perform the operation in real-time. (The worst case is without spirals in the image).

In order to know which distances can occur in the image a histogram of the (5,7) series of distances from the distance image can be made. All non zero distances from the (5,7) series can then be programmed in the successive LUTs.
Another method is to first make a normal skeleton. A rough measure of the thickest object in the image can be found by interrogating the 'stable image' status bits of the CLPEs. All distances from the (5,7) series below this distance can be programmed in the successive LUTs.
The cost of several basic grey value operations is summarized in table 5.5.2

Operation class	Examples and remarks	Stages / GVSs
Monadic & Dyadic Point Operations	Sqrt, Ln, Add, Sub, Mul, ...	1
Recursive form & row & column separable	Unif, Sobel	4
Row & Column Separable filters	Gauss, Lmax, Lmin, Lrank,..	5
Separable Coefficient Schemes	5 point Laplacian	6
General Convolution		9
Histogram	128 values (+ host communication)	16
(Constrained) Distance transform	intermediate memory	16
Pseudo euclidian skeleton	diameter of largest object d	16+d

Table 5.5.2 Number of grey-value slices for some operation classes.

5.5.3 Design variants.

An **all WLA approach** is obtained when instead of a LUT for the word wide point processing, a Writable Logic Array can be used to implement all non global monadic and dyadic point operations. A WLA is beneficial in the sense that it can be loaded in a few clock cycles, whereas a LUT need several thousands of clock cycles. A 16 x 16 WLA can be used to make any possible connection (dyadic Boolean operation) between any output bit and any two input bits of the WLA. However, still an external ripple carry circuit is necessary. Writing out the ripple carry for a *b* bit addition in canonical (and/or) form for the WLA costs:

$$(\sum_{i=1}^{b} i) \text{ OR terms.} \tag{5.5.9}$$

With an additional 36 OR terms for an 8 bits carry vector the WLA size grows to 52 terms and still needs the input WLA of the $CLPE_2$ to perform the second half

addition. As it still is not possible to process global operations like trigonometric and logarithmic operations using the WLA approach, the conclusion must be that provisionally Look-Up table processing is preferable in PipeLine processing. The LUT should be doubled or tripled in size if pre-storage and fast selection of operations is necessary.
Note that the feasibility of a LUT is different for Linear Processor Arrays and Square Processor Arrays, where the number of PEs is a factor of 10 to 5000 higher than in a PipeLine and Look-Up Tables of such a size for processing are hardly applicable.

High precision processing.
As the storage of temporarily images is different at Pipelines and Processor Arrays the solutions for processing images with a word precision higher than the word precision of the pipeline basically differs between PLs and PAs. Whereas the PA can store a carry bit in a separate bitplane, the PL has to store the carry bit in its own form of temporarily storage, a data channel. The PA will process the carry bitplane in a next pass, the PL will process the carry in the next PE. Figure 5.5.6. shows this skip over effect. One of the assumptions is that the PL has enough memory, i.e. data channels, to perform such a procedure on each level.

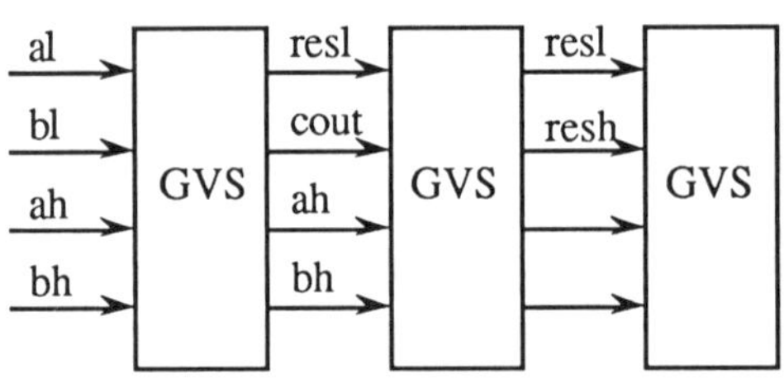

Figure 5.5.6 High precision processing in a pipeline.

The alternative (Duin and Jonker 1988, Jonker et al. 1989) to provide each grey value slice with its own n*n*b bit image memory is highly inefficient and more or less in contradiction with the data flow principle of the pipeline.

Mixed grey value and cellular logic processing.
To enhance the PipeLine efficiency one of the inputs of the CLPE's input WLA is the *fb* or feed_back line. See figure 5.5.3. This line runs from the last CLPE in the Pipeline to all CLPEs one bit lower in significance. In this way it is possible to make

a highly efficient use of the CLPEs both for grey value processing as well as for Cellular Logic Processing. Moreover it enables a flexible bit precision for grey value processing as well. Figure 5.5.7 shows a set-up in which four of the sixteen slices are used in 8 bit precision, two slices are used in 5 bit precision and the rest of the CLPEs (86) can be used for Cellular Logic Operations.

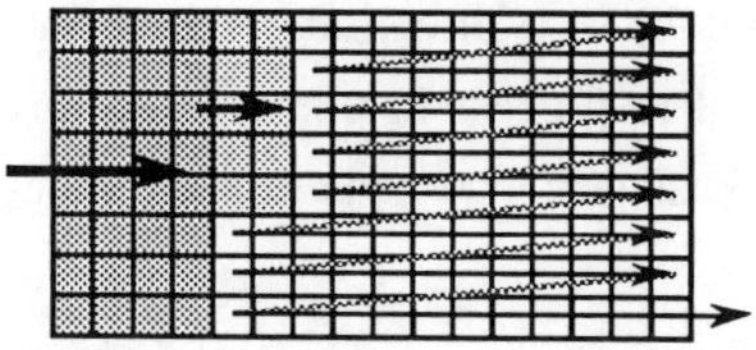

Figure 5.5.7 A flexible precision pipeline.

Local autonomy.

The Greyvalue slice of figure 5.5.3 shows 3 bits from the LUT that are global to all CLPEs in the slice. These bits can be used to program some local autonomy in the slice. Decisions can be made on the outcome of the LUT operation, such as swapping of data. Moreover, control can be passed to the next slices using one of the auxiliary channels. This local autonomy can be compared with if-statements based on the data. Note, that as one of the benefits of pipeline processing is the unfolding of iteration loops, while and repeat statements can only be implemented as local control features by programming a maximum number of iterations and skipping the next PEs if the programmed condition is true. This condition should be passed along through all PEs involved in the unfolded loop and hence a channel (one bit only) should be reserved for this condition. Of course, the controlling host processor can always abort the processing due to some condition, re-instruct the PipeLine and restart the processing. In WLA based systems this can be performed within a single video frame time. This is generally not possible in LUT based systems, when the LUTs should be down loaded.

Host and image memory connection.

Finally figure 5.5.8 shows a PipeLine system based on grey value slices. Input to the system can take place directly from the sensor or from image memory, output to the host can take place via image memory or through status information of the CLPEs at the end of the PL. In long PLs, early return should be made possible for short algorithms. This is quite naturally implemented for Cellular Logic Programming as early return is always possible at the end of the pipeline and further use of the feed-back lines is omitted. An extra tap to image memory can be made on some points in very long PLs to enhance the efficiency for grey value processing. The actual use of the tap can be made programmable. The intermediate memory can be used for the distance transform family of operations. Moreover, as each CLPE has the capability to address image memory in two directions and as for iterative Cellular Logic Operations it is beneficial to process in upward and downward passes, a higher efficiency can be obtained using an intermediate frame storage.

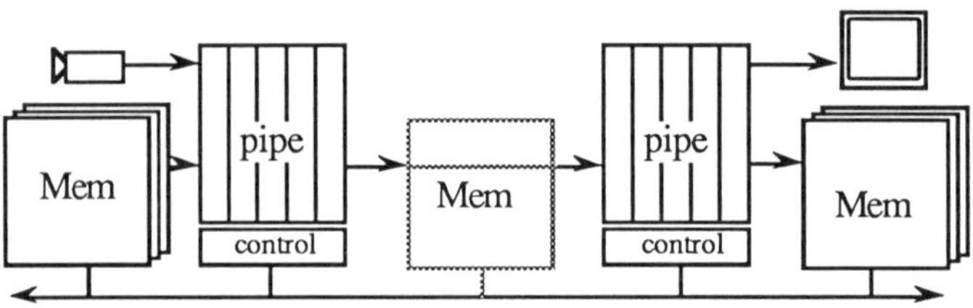

Figure 5.5.8 A Greyvalue PipeLine.

For direct camera input a non interlaced input is assumed. As the line and frame syncs are delayed throughout the pipe, a monitor connection can be made using the sync signals at the end of the pipeline. Moreover, for visual testing of intermediate results sync signals derived from any CLPE can be used.

5.5.4 Conclusions.

It is feasible to build a greyvalue Pipeline system with CLPEs.

There are two alternatives to perform this:

Firstly by incorporating LUT modules in front of a stack of CLPEs and secondly by performing a leap-frog method using a linear string of CLPEs.

The LUT method gives more speed and flexibility in the sense that many point operations can be performed in a single clockcycle. Moreover, it allows data dependent actions to a limited extent. Greyvalue neighbourhood operations can be performed through neighbourhood generation by the stack of CLPEs.

Cellular logic processing remains possible by the introduction of feed-back lines.

With a modest number of CLPEs, a system can be made that is suitable to perform most common low-level operations in real-time.

6 Toward an architecture for low- and intermediate level 2D and 3D image processing.

In this chapter an architecture is discussed that is capable to handle image processing problems for two dimensional and three dimensional images in the field of low- and intermediate level image processing.
Although this research is still continuing, the considerations that has led to the proposed and partially simulated architecture[1] will be treated in the following sections.

Section 1 describes design aspects special to 3D image processing.
Section 2 describes design aspects special to intermediate level image processing.
Section 3 describes general design considerations and a list of system requirements.

Section 4 describes the draft architecture and its constituting parts.

Section 5 describes the implementation of low-level 2D operations on the architecture.
Section 6 describes the implementation of low-level 3D operations on the architecture.
Section 7 discusses the implementation of intermediate level algorithms.

Section 8 contains the conclusions.

6.1 Design aspects of a 3D image processing architecture.

Image processing using three dimensional images usually find its application in biomedical image analysis, although multi-layer PCB and VLSI-mask checking are examples of applications in another field.
Confocal Scanning Laser Microscopy (CSLM) using fluorescently labeled cells, labeled cell nuclei or labeled DNA within the cell nucleus is an example of an application in the biomedical image analysis using 3-D image processing.

[1] In APL-II, according to (Blaauw 1976) and the method discussed in Section 5.3.

The aim of this analysis is the measurement of the position of the fluorescently labelled structure and its intensity. In most 3D imaging methods like CSLM, fewer samples are usually acquired in Z direction then in X and Y direction. If an optimal sampling density is assumed in all three axis directions then basically two options are available to process the 3D images. Using resampling to obtain an equal resolution in each direction, or using image processing routines that account for unequal sampling frequencies. A reasonable assumption is that the frequency in X and Y direction is equal and the Z direction has a much lower frequency, yielding typically 3D images of sizes 256 x 256 x 16, 128 x 128 x 32 or 128 x 128 x 64. However, as the CCD chips that are applied in the CSL microscopes may easily have a X,Y sampling capacity of 384 x 576 to 1300 x 1100 (non interlaced), the effectively chosen 3D image formats are more a compromise induced by limited available memory and the wish for reasonable processing speed than based on the actual CCD size. In this thesis the assumption is made that the sampling rate in the Z direction is less than the sampling rate in X and Y direction and that the sampling rate in X and Y direction is equal.

The acquisition process for 3D images is usually not very fast in comparison with 2D image acquisition in e.g. industrial applications and robotics. This is caused, for example in a CSLM by the required integration time of the electro-optic sensor. The storage order in memory is usually first X (voxels) then Y (lines), then Z (slices). The voxel accuracy may be up to 12 bits, sometimes scaled to 8 bits in one of the first image processing steps.

The 3D image processing routines that are used are routines that are extended from 2D image processing, usually based on Linear and Non Linear window operations and voxel counting. As in many 2D applications, after edge detection the voxel domain is left and further processing is done in the geometrical domain. As a consequence of the non-uniform resolution in X, Y and Z direction, algorithms that are based on the position of the voxel in the image, like the 3D vector distance transform (Danielson 1980, Yamada 1984) the voxel's position (address) is necessary for the vector calculations.

For low level image processing operations in 3D, the speed requirements are induced by the massive number of voxels and not by the high speed of the task. Most image processing algorithms remain the same, with one dimension extra. However, due to the non-uniform resolution, address calculations are more frequently incorporated in the kernel operations then in the 2D case. The possibility to perform complex arithmetic operations on the values and addresses of the voxels in the kernel is a requirement for a 3D image processor architecture.

In order to perform fast cellular logic operations in 3D, several approaches are possible. A plain table look-up in a 3 x 3 x 3 kernel is impractical, although resampling the image from a cubic tessellation to a tetradekahedral tessellation (the 3D extension of the 2D hexagonal tessellation) and using a 13 element kernel, makes hardware based on a 8192 location look-up table possible (Preston 1991). The disadvantage of this scheme is the same disadvantage as the 2D hexagonal tessellation and lays in the necessity to resample the square sampled image. A scheme based on the sequence of 2 x 2 x 2 table look-up (Lobregt et.al. 1980) has as a disadvantage that the procedure requires a sequence of look-up actions and some additions. The scheme based on mask sets, as was introduced in sections 4.1 and 4.2 allows the processing of a voxel in a single clock pulse, provided that the 3 x 3 x 3 neighbourhood can be accessed in parallel. However, a disadvantage of this method is the vast amount of WLA terms needed (See section 4.2 and Appendix A). It was found that mirroring, rotating and reducing the 3D masks (of Appendix A), resulted in the following numbers of masks for the 3D skeleton with endpoint condition:

- 1 Erosion mask
- 128 Curved surfaces breakvoxel masks.
- 192 Space curves breakvoxel masks.
- 1 Single Point (breakvoxel) mask.
- 82 Curved surfaces edge masks.
- 24 Space curve line endpoint masks.

Resulting in a mask set of 428 masks for the 3D skeleton with endpoint condition and a mask set of 321 for a skeleton without endpoint conditions.

Using these masks with the cascadable version of the Writable Logic Array (section 5.1) and a parallel 3 x 3 x 3 neighbourhood update, a 3D Cellular Logic Pipeline can be constructed. Note that 3 x 3 x 3 inputs constitute the normal neighbourhood, 3 x 3 + 4 inputs constitute the recursive neighbourhood, 1 input is required for mask image Z and 6 inputs are necessary for an opcode to group mask sets. A total requirement of 47 inputs. A feasible WLA size at this moment, however, is a WLA of 20 inputs by 32 terms (about the CLPE-WLA size). Then a cascade of 3 WLAs -to enlarge the number of inputs-, by 16 cascades in parallel -to enlarge the number of terms-, is necessary to form a single Processing Element for a 3D Cellular Logic Pipeline. This results in a total of 48 WLAs. Note that a cascade of 3 WLAs needs 3 clockpulses to process all inputs, but even this can be pipelined resulting in still a single clock-pulse per voxel.

6.2 Design aspects of an intermediate level image processing architecture.

As was concluded in chapter 1, the field of intermediate level image processing algorithms is large and vague. An important characteristic, however, is that within this class of algorithms, a transformation takes place from the image data structure to some other data structure. In order to understand what this implies for a new architecture and to develop primitive functions in hardware that can be used to speed up these algorithms, the A* transform and the region growing problem as described in section 2.3 have been investigated as example problems to be solved by the new architecture.

6.2.1 Considerations on a special architecture for the A* algorithm.

The A* algorithm was briefly described in section 2.3.3.
To obtain a feasible hardware architecture, several implementations of A* have been compared, both theoretically and in simulations (Dekker 1987, Jonker et al. 1988a, Verwer et al 1988b) and are summarized below:
Due to the evaluation of the functional simulations it was decided to focus on an architecture for the A* algorithm using bi-directional search, directional a-priori pruning, a-postiori successor pruning and (optional) heuristics.
An obvious source of parallelism in the A* algorithm is the wave front of equal cost that has to be expanded. It was decided that the path finding machine (PFM) should consist of identical processing elements, in which each PE handles a number of 'open' nodes. Moreover, the PEs should only have access to a mutually disjoint subset of the large data structures or images. We will elaborate on this later. Each PE can access the data on only a subset of the nodes in the search graph. Communication between the PEs remains necessary to exchange successor nodes. Figure 6.1 shows

the architecture of the PFM consisting of a number of PEs, a data switch network and a single global controller .

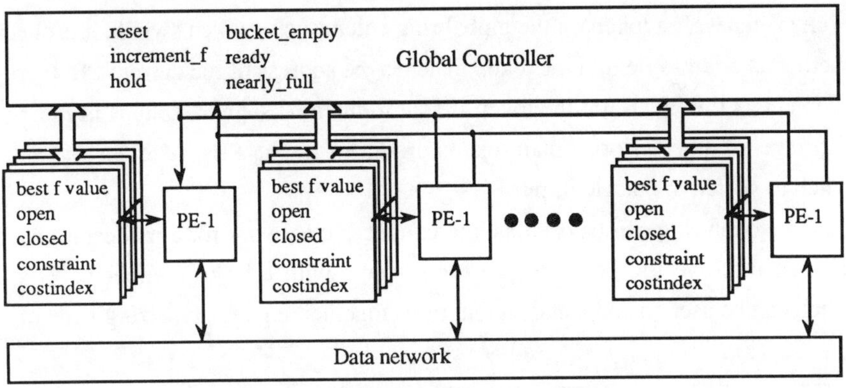

Figure 6.1 The architecture of a Path Finding Machine.

The global controller synchronizes the PEs and might be implemented with a uniprocessor. The input and outputs of a PE are *reset, increment_f, bucket_empty, ready* and *nearly_full*. During the actual search process, the global controller has only a passive job, which can even be implemented with some logic. If *reset* is asserted, all PEs clear their current evaluation value f. After reset, each PE starts to process the nodes in the bucket indicated by their local f value. When a PE is ready with this bucket, it asserts the *bucket_empty* signal. The PEs may however proceed with the next buckets, (f+1), (f+2), ... until the bucket number exceeds f+d; d being the minimal transition cost, e.g. 5. (Minimum of {5,7}; see also figure 2.3). Note that proceeding is allowed since buckets with index f generate successors with an evaluation value of at least f+d. The global controller waits until all PEs assert the *bucket_empty* signal. At that moment, *increment_f* is asserted and all PEs are enabled to proceed. If some PE detects the end condition, it sends the *ready* signal to the global controller, which then terminates the search. The global controller's active tasks are initialization and finding the final solution path.

The data network transfers successor node data between PEs and can be implemented for a modest number of PEs with a data bus. If the PFM consists of more than a modest number of PEs, a bus will become the engines bottleneck. A variant of the token ring approach as used in the Manchester data flow system (Gurd et al. 1985) and the IMage Pipeline Processor (Iwashita 1986) can be used. See also section 3.5.5.

The PEs are connected in a local ring, see figure 6.2. The elements labeled 'S' are token switches. If a token on the global ring enters such a token switch, it is checked whether an address field in the token matches the address of the current PE. If so, the token enters the PE; if not the token is sent to the next switch. Tokens in the global ring have a higher priority than tokens produced by the PEs. A token walks from switch to switch in one clock-pulse.

A single token ring can be used as interconnecting network for a modest number of PEs. To improve the performance for systems with a large number of PEs, the switch can be used to connect different rings together, e.g in a hyperring structure.

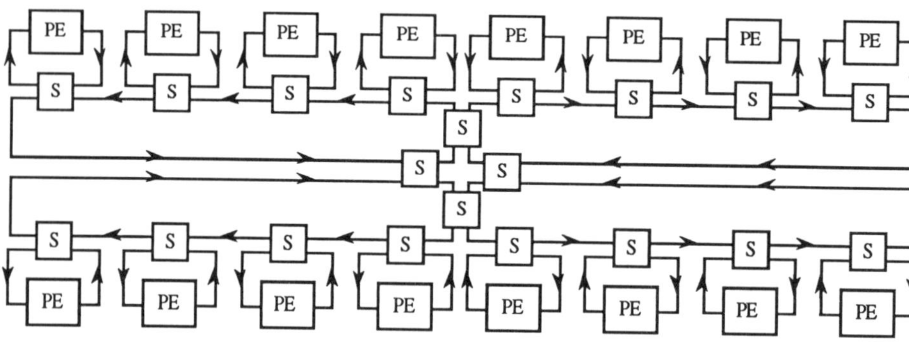

Figure 6.2 Connection of PEs in a (hyper) ring structure.

A draw back of this architecture is that parallelism cannot be forced; but relies on the uniform distribution of the nodes over the PEs. This distribution of the search graph nodes (the wave front) is a fixed, fine grained, spatial distribution, and is implicitly executed by the processor mapping function (PMF). This function assigns a PE number to a coordinate. A simple mapping function is based on the binary representation of the coordinates: The least significant bits of the coordinates are used as PE address. Simulations (Dekker 1987, Jonker et. al. 1988a) showed that the total number of input nodes in 'open' was divided almost uniformly over the PEs, with as notable exception one of the test images; a generated 3D maze, which was too regular to achieve an even distribution. See also Appendix C for the test-images that were used in the simulations.

Figure 6.3 shows the architecture of a *processing element*. Each PE is a synchronous pipeline that handles, enabled by the global controller, a number of open nodes. Each clock cycle every PE generates a successor node, which is either stored in its own input FIFO or over the network in the FIFO of another PE. The PE selection is performed by using a few bits of the node's coordinates as PE address.

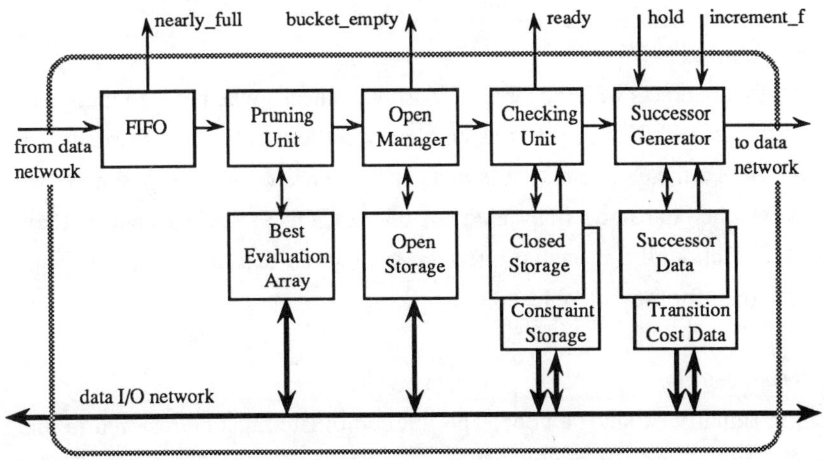

Figure 6.3 The architecture of a PFM Processing Element.

The PE consist of a number of pipelined functional units, each with memory for data (images) and parameters. The individual tasks of those units are listed below:

Input fifo:

On the average, the OPEN manager processes input nodes at the same speed as the successor generator creates them. However, the momentary node rate can be higher, since at times successors from different PEs can all be sent to the same PE. The FIFO generates the *nearly_full* signal, which can be used to temporarily hold the successor generators of other PEs.

Pruning unit:

Uses a 'best evaluation array' or 'distance image', which keeps for every node coordinate the best evaluation value so far. Only nodes that are 'better' than the best evaluation value so far may pass. The array can also be used as traversal tree by the global controller to find the final solution path.

Open manager:

The administrator of the local 'open' data structure, organized in buckets. Input nodes are stored in the bucket indicated by their f value. Output nodes are retrieved from bucket $\{f_x \mid (f \leq f_x < f+d)\}$. The *bucket_empty* signal is generated by this block.

Checking unit:

The unit first interacts with the CLOSED array. It checks whether the input node is an element of 'closed'. If the node was an element of 'closed', it is discarded. If not, it checks whether the end condition is met. Then the *ready* signal is asserted. If not, it checks whether the node is an element of the CONSTRAINT image; then it is discarded. Optionally, it has an interface with some external constraint checking hardware (e.g. for moving objects).

Successor generator:

This unit actually expands a node. The full coordinate must be restored in this unit from the local coordinate and the PE number. Depending on a direction field of the node, successors can be generated from a table. This table can be filled with certain types of successors, such as successors only in forward line with directions found in the generating node or under obtuse angles with this direction. The new evaluation value is generated for each successor. Optionally, a cost index storage can be connected containing a cost index for each local node coordinate. (To allow space dependency of costs, transitions near objects can be given higher costs). The successor generator has a heuristic unit to calculate the heuristic function h.

For the register transfer simulation of the PFM, each of the units was further refined and the following assumptions were made:
All units of the PE are separated with pipeline registers and a two phase clock is used: ($\O_1$; compute and $\O_2$; transfer). All units could be constructed in such a way that the processing time needed for a computation step is twice the cycle time of the RAMS used for the tables. Hence a realistic cycle time[2] is 200ns. All units take one elementary cycle, except for the successor generator which takes one cycle per

---

[2] The original realistic timing figures (1987) were adapted to current realistic timing figures (1992), (divided by two).

generated successor. If a heuristic function is used, two cycles per successor are necessary.

The simulations have been run for 8 test images {a...h} using all combinations of unidirectional/bi-directional search, pruning/no pruning and heuristics/no heuristics (see also Appendix C). A-priori pruning was not applied, since the benefits of this have been discovered later on. The distance coefficients used (transition cost d) were 5, 7 for 2D images and 7, 10, 12 for 3D images. Table 6.1 shows the data of a typical simulation.

#PEs	clock cycles	PE	FIFO depth	open nodes	checking unit			successor generator
					idle	blocked	nodes	idle cycles
1	275129	0	1	200644	4	77363	41030	72213
2	126027	0	13	99865	4	27148	20507	40404
		1	4	84608	15	43035	20396	25081
8	30146	0	11	22376	312	7714	5124	5655
		1	16	21480	313	8848	5068	5336
		2	10	21943	208	8271	5143	5504
		3	13	21748	139	8722	5089	5237
		4	8	23889	2313	4130	5117	10189
		5	12	23485	964	5977	5109	7692
		6	9	24443	1824	4088	5130	8954
		7	12	23851	549	6032	5129	7334

Table 6.1 Typical simulation output for the PFM: For a 1, 2 and 8 PE system, the number of cycles, the FIFO depth, the number of nodes in 'open', the number of nodes after checking, the idle- and blocked cycles of the checking unit and the idle cycles of the successor generator are given.

Table 6.2 gives the processing times of the path finding machine, using no heuristics, no a postiori pruning and bi-directional search.

image	1 PE	16 PEs	speed up
a	66.0 ms	4.8 ms	13.8
b	55.1 ms	3.4 ms	16.2
c	79.2 ms	4.6 ms	17.2
d	51.8 ms	3.9 ms	13.3
e	96.1 ms	5.1 ms	18.8
f	123.3 ms	7.0 ms	17.6
g	8.8 ms	0.6 ms	14.7
h	4.9 ms	0.8 ms	6.1

Table 6.2 Processing times of a simulated PFM. The processing times are based on clock cycles of 200 ns.

The software version (in C) of the algorithm using image e, runs in 7 sec. on a 40 MHz uni-processor and the PFM takes 5.1 msec., a speed improvement of 1300.

Discussion:

The machine that was designed can be seen as a Square Processor Array with an unusual mapping of its PEs over the image. See figure 6.4a. Moreover, all PEs not only have nearest neighbour connections, but are connected with all other PEs, although the connection speed varies due to the logical distance of the PEs in the hyper-ring and the contents of the input FIFO queues. Due to this processor mapping function (PMF) all elements on a processing front are more or less uniformly distributed over the PEs.

1	2	3	4	5	6	1	2	3
4	5	6	1	2	3	4	5	6
1	2	3	4	5	6	1	2	3
4	5	6	1	2	3	4	5	6
1	2	3	4	5	6	1	2	3
4	5	6	1	2	3	4	5	6
1	2	3	4	5	6	1	2	3
4	5	6	1	2	3	4	5	6

a)

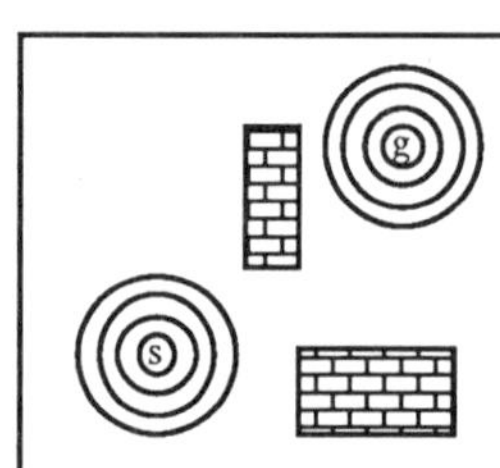

b)

Figure 6.4 a) A Processor Mapping Function. b) A wave front pattern.

Consequently, all PEs are involved in the processing of both the start and the goal wave front. See figure 6.4b.

Breadth first search (A* without heuristic function h) with distance coefficients $\{5, 7\}$, is totally equivalent to the $\{5, 7\}$ Distance Transform, and consequently also a Recursive Neighbourhood Operation. The difference is, however, that a write formalism is used instead of a read-formalism.
The read formalism is a custom in most current low-level architectures: The central pixel reads the neighbourhood values from his neighbours, performs the neighbourhood operation and stores the result.
With the write formalism, all PEs send their data to their neighbours and all PEs accept data from their neighbours in order of reception of the data from their neighbours. The difference is little, apart from the fact that now both the initiative and the order is determined by the neighbours from the neighbourhood instead of by the central pixel itself. The PE processes the data until the entire neighbourhood is processed and writes the result back, i.e. closes the node. It will be clear that a bookkeeping has to be kept for the write formalism. If a neighbourhood value is first entered for a node (pixel) the node is opened, if the last neighbourhood value has arrived the node is closed.
Due to the fact that there are many more nodes (or pixels) than PEs, each PE can simultaneously (but sequentially) process neighbourhoods of many open nodes.
The synchronization still comes from a global controller and a local priority to process certain nodes before other nodes is given by a possible (bucket) sorting property of the PE and the priority setting of the global controller.
This architecture can be labeled a Virtual-SPA and the processing can be (with some skepticism) labelled SIMD with local autonomy, as all PEs perform the same process, synchronized and instructed by the global controller.
The benefit of this architecture is that with a limited number of powerful PEs a powerful full size SPA can be emulated. However, this architecture is only beneficial for operations that take as least as long as the transport time from PE to PE. For straightforward kernel operations involving the entire image, this architecture is not suitable. The fact that the process activity takes place on relatively small fronts in the image, results in a high system efficiency due to a minimum number of idling PEs. (This high efficiency in comparison with a full size SPA.)

6.2.2 Considerations on a special architecture for region growing.

The region growing algorithm was briefly described in section 2.3.4. In this sub-section a theoretical discourse will be held on the subject of region growing algorithms, to derive requirements for an architecture that is capable to support this.
The obvious inherent parallelism in the region growing problem is twofold. Each region can be grown from a seed more or less independent from other regions, and all pixels on the closed contour wave front per region can be treated in parallel. The difference with the A* algorithm, however, is that there is no principle distinction in the wave fronts (e.g. start and goal front) and there are no global parameters to obtain per front.
Assume the seeds are obtained by a series of normal low-level routines, e.g. processed on a low-level architecture. The pixels on the wave front can be processed with front processors connected to a hyperring, equivalent to the A* implementation. This is a comparison action in which the surface parameters of one front are used. These parameters are submitted by plane parameter update processors (see figure 6.5).

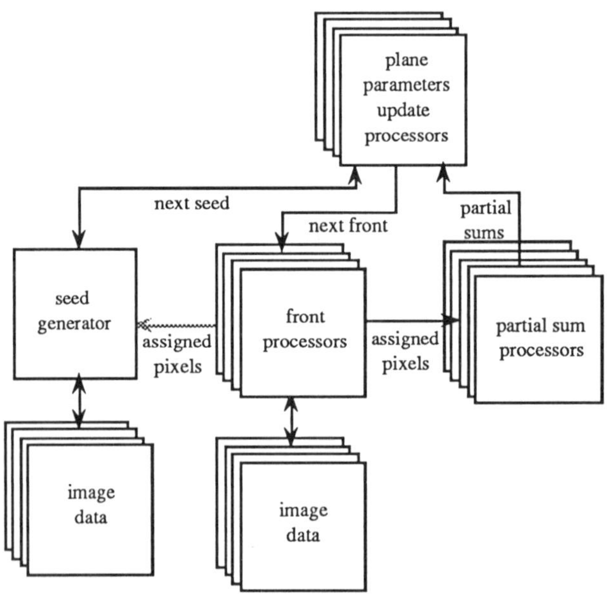

Figure 6.5 Task assignment in the region growing algorithm.

The result of the comparison action of the front processors is that pixels are assigned to the front or rejected. All assigned pixels are sent over the ring to the queues of a second set of processors, the partial sum processors, that maintain the partial sums for fast calculation of the plane parameters.
Optionally, the front processors may pass the pixel assignment to the seed generator, thus preventing that seeds already assigned to a front are passed.
If an entire contour of the front is processed, gradually all processors halt, and the plane parameter update processor(s) calculate the new plane parameters.

Whereas the front processors still operate in the pixel domain, the partial sum processors operate in the object domain. Note, that these processors may be identical to the front processors, only their task is different. However, the required bit precision may be different. For example front processors: 16 bit integer, partial sum: 64 bits integer, parameter update processors: floating point.
Provided with new plane parameters the next contour of the front of the region may be processed. However, there is no objection in a context switch at the moment that an entire front is processed. The contour of a second front may be processed in parallel to the parameter update of the first front. The parameters of the second front can be loaded in the front processors during processing of the first front.
It will be clear that the A* method cannot be used in algorithms where global object data has to be maintained as this would lead to a context switch at every investigated pixel at the front.

In the previous scheme only a single front was processed at the time. If the processors are separated in groups than each processor group (PG) can process a separate wave front in parallel to the other processor groups. See figure 6.6.
Note that as a consequence, for N processor groups, N sets of images should be maintained and groups are able to assign the same pixel to their wave front, so double assignment may take place.

The double assignment problem does not manifest itself when sequentially processing wave fronts, as first come first served is implicit to this approach.
In parallel processing of the fronts, arbiter processors monitoring the assignments, have to judge which assignment is better. They can perform this task by comparing the errors in both assignments. As a wave front contour is finished, an arbiter processor can adjust the contours. Due to the necessary front revision period, the

system performance is improved if PGs can also work on different fronts, to allow adjustment by the arbiters.

The problem that two PGs are actually processing the same region due to wrong seed composition can partially be overcome by passing the assignments to the seed generating processor. In all other situation, usually the fuzzy areas around contact contours of regions are subject to the judgement of the arbiters. Suppressing of spurious branches onto surfaces to avoid communal assignment, can also be taken by applying extra local assignment rules (hence performed in the front processors), based on the number of neighbours that are already assigned in a certain neighbourhood. They diminish the workload of the arbiters, but do not prevent their necessity.

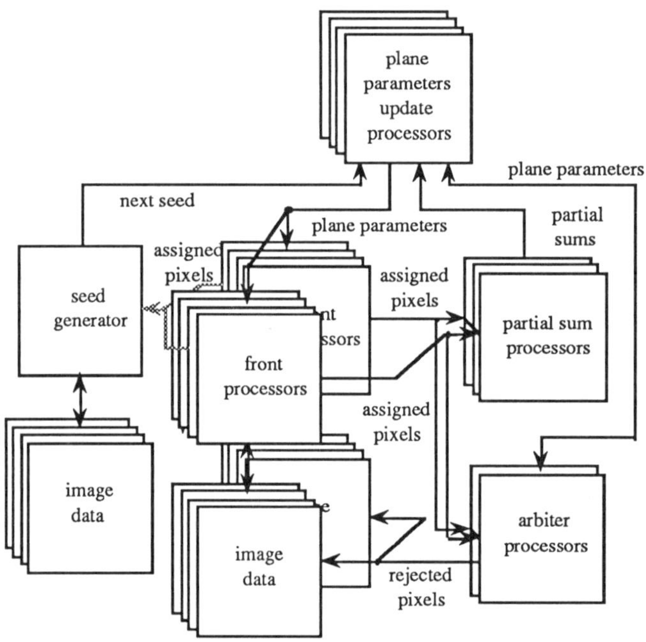

Figure 6.6 Task assignment to processor groups in the region growing algorithm.

Discussion:

From the A* implementation we can conclude that: For more complex algorithms in a Local or Recursive Neighbourhood, focussed on objects or contours of objects, concentration of the PEs around these activity centers can be obtained by assigning

the pixels of the image with a Processor Mapping Function to the PEs, rather than making a fixed assignment by the data network as with the SPA. The PEs can be more powerful because less PEs are needed. The PEs should be connected with a suitable and fast data network in which they can address each other.
Although the PEs may be identical, groups of PEs may be clustered to PE groups (PGs) whereas PGs may be assigned to identical parallel tasks, but also to different tasks. Different types of PEs may also be used if connected with the same data network. It can be identified that PGs can be assigned to wave front maintenance tasks, to object maintenance tasks, to object interaction tasks and to global scene tasks.

6.3 General design considerations.

In this sub-section, general design considerations and a list of requirements for a proposed architecture for 2D and 3D low- and intermediate level processing is provided.

In (Komen 1990) the differences between four basic groups of architectures for two dimensional low-level image processing are discussed. In section 5.2 this discussion was extended with an emphasis on real-time image processing. The conclusions that can be derived from these discussions are that:

SIMD Square Processor Arrays suffer mostly from a lack of sufficient number of PEs as well as a wiring problem. As soon as the image size exceeds the array size, some form of scanning with the array over the image must be performed, which makes the programming of the machine cumbersome and gives a loss in speed due to the scanning overhead involved, typically a factor of 4.
The wiring problem reveals itself by the sequential solutions often sought for loading and unloading data in and out of the array and loading instructions into the array. The instruction word size (mostly larger than 8 bits), usually means that a sequential solution is sought for loading the instructions into the PEs. If sufficient instruction pipelining in the PE can be performed this may not cause any delay. However, when data dependent branching occurs on array data, instruction caching cannot be used and apart from the delay caused by the controller instructions, the delay caused by the serial instruction input may press on the performance. If the SPA has a sufficient

number of PEs, its speed will be high enough for real-time processing. However, when a certain limit is exceeded, the frequency of operation halves, unless twice as many PEs are taken to solve this problem.
SIMD Linear Processor Arrays have similar controller and instruction cache stall problems as the Square Processor Arrays (see section 5.2). However, due to the limited number of PEs and the linear configuration, parallel instruction input is more likely and hence not a bottleneck any more. LPAs usually have no data I/O problems. The LPAs speed will usually be high enough for real-time processing, although due to the limited number of PEs in comparison with the SPA the limit above which real-time operation is not longer possible is reached sooner. Here also the frequency of operation halves, unless twice as many PEs are used to solve this problem.

A Pipeline has no controller overhead at all, due to the fact that reaction on results of operations can usually not be performed, so data dependent branching is hardly possible. The controller programs the pipeline for one entire task. This is a disadvantage. One of the advantages of the Pipeline is that even with a few PEs, image processing can be performed and that adding a small number of PEs may be sufficient to enhance the performance of the pipeline, if the real time performance is endangered.

A few observations can now be made:
Cellular Logic Processing can be performed in one clockcycle per pixel, due to the fact that the entire neighbourhood can be accessed and processed instantaneously, e.g. using a look-up table. For greyvalue operations this is impossible, due to the fact that most greyvalue operations are based on a dyadic integer operation. Consequently all pixels of the neighbourhood need to be accessed in a sequence, due to this operation. On the other hand, row and column separation of kernels, the use of partial sums, sparse kernels and kernel sizes larger than 3 x 3 are more likely to occur in greyvalue processing than in Cellular Logic Processing. Kernel sizes larger than 3 x 3 are not as likely for Cellular Logic Operations.
Large shift registers for neighbourhood creation are only technically feasible[3] when they are bit-wide and not byte-wide. Real shift registers make the parallel tapping of the neighbourhood possible, in contrast with shiftregisters based on RAM. Flexibility

[3] To allow the pipelined processing of 3D images, entire 2D images have to be stored in the shiftregisters, this is currently unfeasible.

in the neighbourhood acquisition can on the other hand easily be achieved with RAM and a proper addressing unit and not easily with shift registers. This points to the direction that Pipelines can be be built effectively using shift registers and Linear Processor Arrays by using RAM.

Section 4.2 showed that for Cellular Logic Operations a single clock pulse per pixel is theoretically possible even for dimensions higher than 2D, if a WLA approach is used in a parallel acquired recursive neighbourhood.

These observations point to a combined Pipeline solution with parallel neighbourhood acquisition for Cellular Logic Operations and a SIMD Linear Processor Array solution for greyvalue operations. For spatial recursive greyvalue operations like the distance transform, however, sequential and hence pipelined processing is also beneficial.

Due to the video standard, image data I/O can be performed with separate shift registers that can shift in and out entire video lines.

However, as soon as the processing activity in the images concentrates on more complex operations on separate structures or regions in the image, virtual SPA processing can be obtained by processing with a write paradigm instead of a read paradigm, using PEs with (bucket sorting) queues, and an appropriate autonomous data network through which PEs can communicate by using data as PE addresses. The assignment of PEs to pixels is performed by the Processor Mapping Function.

Parallel processing of the separate structures in the image can be performed by clustering PEs in processor groups (PGs), each with their own PMF, with as a possible consequence that two PEs patronize the same image element. PGs can be assigned to wave front oriented tasks (pixel level), to blob oriented tasks or to global tasks. The word length and the capability of the PEs in these task groups may vary.

These considerations leads to a preparatory list of requirements that can be used for the design of a 2- and 3D low- and intermediate level Image Processor (23DIP):

The 23DIP should ....

a) support 3D as well as 2D low-level image processing operations, both in the greyvalue and in the cellular logic domain.
b) be able to process 2D images of at least 512 x 512 pixels for 2D image processing purposes and up to 16K x 16K pixels for VLSI and PCB mask

checking. It should be able to process 3D images of at least 256 x 256 x 64. Preferably this size should be easily expandable.

c) have functions for cellular logic operations that are comparable with the CLPE, including its extension to 3D.
d) be able to function as a Pipeline system for recursive operations, both cellular logic and greyvalue.
e) be able to function as a Linear Processor Array for non recursive cellular logic and greyvalue operations.
f) allow flexible neighbourhood sizes and neighbourhood update techniques for greyvalue operations.
g) be usable in a real-time environment for 2D images.
h) have fast image transposing capabilities in 2D and 3D.
i) function as a system with only a few PEs to a massively number of PEs.
j) will have a nearest neighbour network as well as a fast mail network to other PEs.
j) allow different types of PEs to be attached in the system.

6.4 Draft architecture and description of constituting parts.

A draft system design of the 23DIP is depicted in figure 6.6. This system is partially simulated[1] in APL-II.

The system is built around a linear array of identical PEs, with nearest neighbour connections (data and control). The control connections are meant to synchronize PE operations in their different operation modes. Apart from the nearest neighbour connection a global bus is used. The global bus is used for broadcast purposes. A synchronous ring interface can be used for wave front processing in Processor Groups. The Host/data-I/O interface can be used for high speed loading of the image memories of the PEs.

Each PE is capable to routinously process control sequences, independently from - but in synchronization with- other PEs. This allows the PEs to either process in a 2D or 3D pipeline configuration, to autonomously process entire 2D and 3D images in a Linear Processor Array configuration, or to process as part of a Processor Group one or more wave fronts. In order to allow these modes of operation, a separate downloadable control unit is used (PE_c), that controls the sequencing of the operation modes.

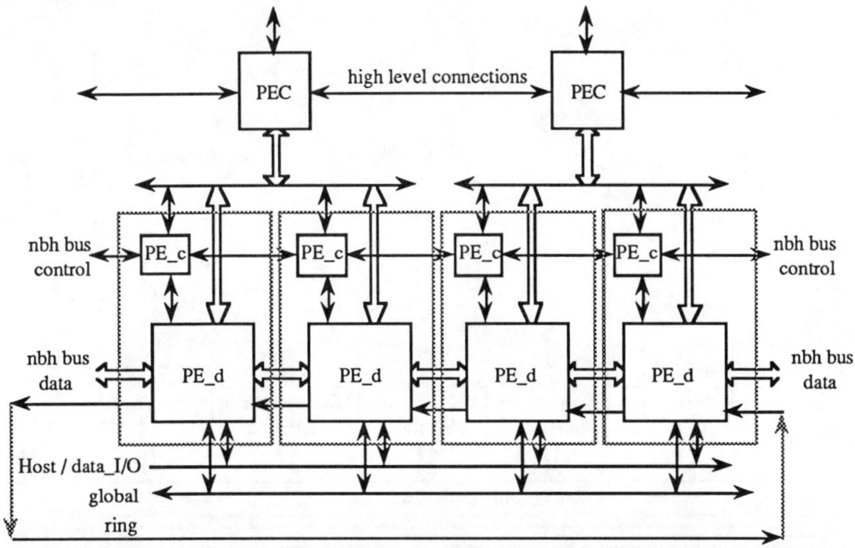

Figure 6.6 Draft architecture of a 2&3D-Image Processor.

Groups of PEs are controlled by a PE Controller (PEC), which itself is a PE from a MIMD system, having its own interconnection network. The PECs control the PEs through their system data and address bus.

The control strategy is that all settings are downloaded from the PECs into the PEs, after which the PEs are allowed to proceed autonomously, allowing the MIMD system to proceed with other tasks. If necessary (non-standard) control sequences for the PE_c can also be downloaded from the PEC.

Figure 6.7 shows the draft PE architecture (datapath). The PE is grouped around six busses of 9 bits wide. The busses can be used to transfer data as well as addresses. Each device that is coupled onto the bus can be configured to use certain busses for addressing and / or reading and writing data. The configuration is performed by the PEC. The actual timing of reading and writing using Read (RD), Write (WR) and Output Enable (OE) signals, is performed by the CWLA (=PE_c) controller. To store the configuration, planned and handed out by the PEC, configuration registers are internal to each device.

PEs are normally meant for processing 2D images in LPA or Pipeline mode. PEs can be grouped to process 3D images.

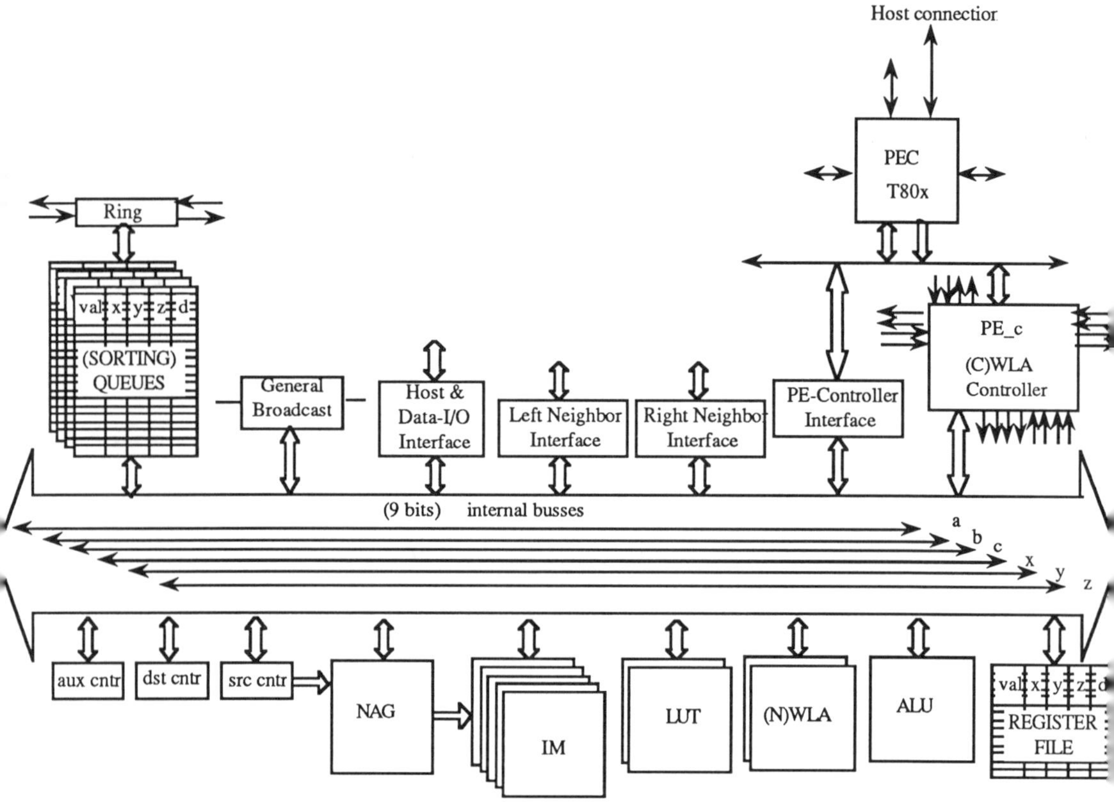

Figure 6.7 Proposed PE architecture.

The devices (datapath blocks c.q. functional units) are:

- Source image counter (X, Y: 2 x 9 bits) to up/down count source addresses for 512^2 images.
- Destination image counter (X, Y: 2 x 9 bits) to independently up/down count destination addresses for 512^2 images.
- Auxiliary counter (X, Y: 2 x 9 bits) to independently up/down count multiples of 512^2 images to process larger images, or to count slices of 3D images.
- The Neighbourhood Address Generator. Separately connected to the source image counter, is able to generate a 3 x 3 neighbourhood in parallel for cellular logic

processing or neighbourhood pixels for greyvalue processing. Any kernel configuration up to 9 pixels can be retrieved without reconfiguration.

- Look-up tables, 18 bits in 18 bits out, for various arithmetic operations on images.
- ALU (32 bit) for more accurate arithmetic operations required in intermediate level algorithms.
- Neighbourhood WLA for cellular logic operations.
- Image memory (512^2 images). Bitplanes are independently addressable by the NAG.
- Register file for the storage of constants and parameters.
- Bucket sorting queues for intermediate level processing. One queue is coupled to the input side of the mail network, one queue is coupled onto the output side of the mail network. Two more bucket queues are for internal use.
- The left and right neighbour interface are tristate buffers allowing the connection of all or some busses of left and / or right neighbour onto each other. This tolerates the addressing of memory by a neighbouring PE.
- The global busses (2 of 6) allow global broadcasting to all PEs or simultaneously status propagation from all PEs.
- The Host & data I/O interface allows fast loading and unloading of image data.
- The PEController interface allows memory mapped access of the PEC onto the PE busses and hence the PE devices, to read and write settings or data for processing purposes.
- The control unit of the PE is a WLA that can be used to download sets of loosely coupled statemachines that handle the control sequences (see also section 5.3.3).
- The PEController is a T80x transputer. The data and address bus of the transputer's memory interface is coupled onto the PE controller interface of the PE. As the PE_c controls the traffic on the six PE busses, the PE_c needs also a separate connection to the PEC bus. The T80x is chosen for its inherent parallel processing paradigm, its MIMD character, its floating point capability and its software support. Moreover, they can mutually be synchronized, to each other and with the PEs.

This architecture was simulated down to a logic level, however, using a device description on the functional level. We refer to this architecture as the 'simulated 32DIP'.
Some devices are detailed below.

The Image Memory (IM):

Figure 6.8 shows the solution for the Image Memory (IM).

The image memory may contain several images (e.g. 4 to 6). Each image has the following structure:

Two data selectors allow reading from and writing into the RAM from the PE busses. Addressing the RAM can either be performed from the NAG source counter combination (*nba*) or from a PE bus pair. The data input can be selected from one of the six busses (*ibusel, obusel*). On negative addresses from the NAG, the edge value from the edge value register is inserted instead of the RAM content.

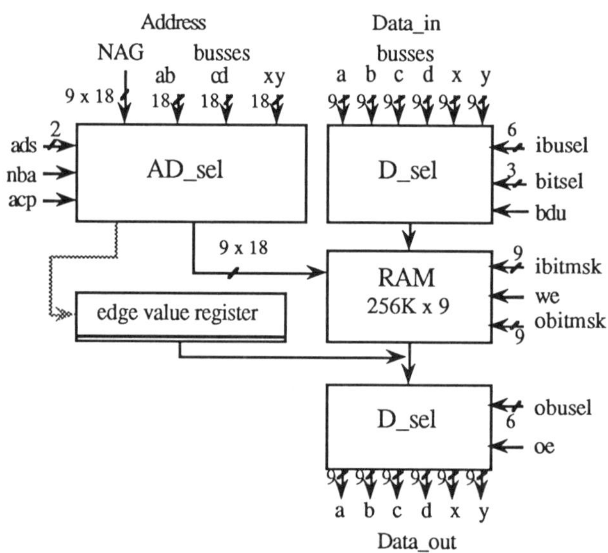

Figure 6.8 The Image Memory (IM).

As the RAM is constituted from 9 parallel 256K x 1 chips, each 512 x 512 bitplane can be addressed by a separated address. The NAG is able to furnish these independent addresses. The idea behind this option is the fast, parallel generation of the 3 x 3 neighbourhood for cellular logic operations. The underlying assumption is that usually the same amount of input, output and temporal images are used in greyvalue processing as are used in cellular logic processing. Consequently, if sufficient storage capacity is assumed to do greyvalue operations, this storage can also store a single bitplane in all bits of a greyvalue image. If a 9 bit image storage is used instead of an 8 bit storage, the entire 3 x 3 neighbourhood can be accessed in a

single clock pulse, if provisions are made to address this neighbourhood and to select and expand specific bitplanes to 9 bit images and vice versa. The signal *bitsel* on the data input selector in figure 6.8 selects one of the bits of the data input and copies this bit to all bitplanes if the signal *bdu* is asserted. Moreover the signal *nba* can be used to copy the central pixel address of the source to all bitplanes, or to use the individual addressing of the bitplanes. Generally, the first occurs in greyvalue processing, the latter in cellular logic processing.
All bits on input and output of the RAM can be masked off (*ibitmsk, obitmsk*).
It is assumed that reading and writing takes one clock-cycle.

The Neighbourhood Address Generator (NAG):
Figure 6.9 shows the set-up of the Neighbourhood Address Generator (NAG), connected to the source counter. The NAG can be used to generate the addresses in the Image Memory of a 3 x 3 neighbourhood as well as the addresses of any arbitrary pixel within a neighbourhood for greyvalue processing. X and Y counters can be presetted with an offset value and are assumed to count a programmable number of bits.

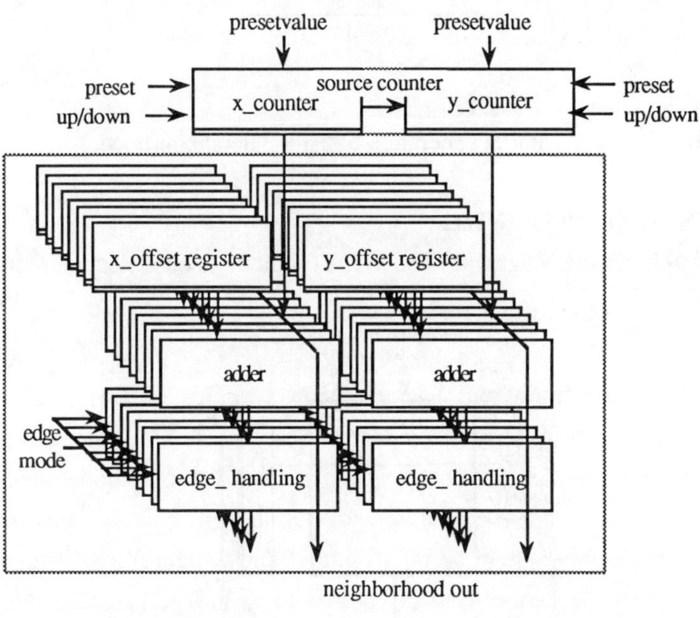

Figure 6.9 Neighbourhood address generator and source counter.

This allows for subimage processing. Within the X and Y offset registers the offset to the central pixel address is stored. This offset for each biplane is added to the central address. Edge handling hardware takes care of either inserting the value of the edge register (via the IM) on addressed image positions beyond the image edge, or mirroring the values of pixels on positions at the same distance from the edge within the image, or repeating the pixel value just before the edge, or taking the pixel value of the wrapped position of the original pixel. It is assumed that XY-counting, neighbourhood pixel address generation and edge handling takes three clock cycles. Figure 6.9 shows the generation of the 3 x 3 neighbourhood by the NAG, using the central pixel's address as input. The NAG idea was adapted from (Zeelen 1986, Zeelen and Jonker 1987) and partially implemented by (Houten 1988).

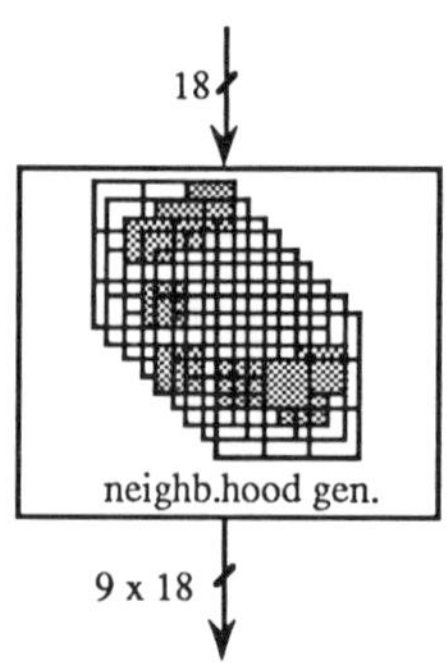

Figure 6.9 The parallel generation of a 3 x 3 neighbourhood.

The Neighbourhood WLA (NWLA):

Figure 6.10 shows the Neighbourhood WLA (without its bus selection units).

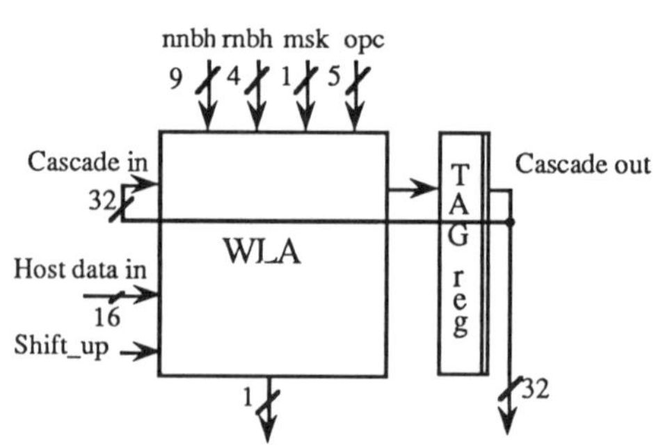

Figure 6.10 The Neighbourhood WLA.

The NWLA is assumed to be a cascadable WLA as introduced in section 5.1 with 20 inputs by 32 terms. This WLA can be used for 2D cellular logic Processing in a single clockpulse.

In section 6.1 it was argued that 48 WLAs are necessary per Processing Element for a 3D Cellular Logic Pipeline. The solution to this large WLA problem, is to cascade three WLAs to account for the large number of inputs and to put 16 WLAs in parallel in a second dimension to cope for the large number of masks. However, this scheme has severe drawbacks. One drawback is the necessity of grouping PEs in groups of three with extra mutual 32 bit connections between them for the WLA cascade. The complex timing associated with the pipelining of the operation is another complication. Due to the less stringent speed requirements in 3D processing, the design decision was made to use a TAG register to store intermediate results, yielding the processing of a 3 x 3 x 3 neighbourhood in three clock cycles instead of one. Note that still 48 WLAs are needed to store the masks.

The TAG register can also be read-out through the busses to locally determine which masks had a hit. This may be useful for e.g. surface measurement purposes in 3D image analysis; the number of hits of a surface in a 3D image on each surface mask may be used to compensate the error in the estimation of the surface area. [Mullikin 1992].

The Sorting Queue Memories (SQM):

Figure 6.11 shows the datapath for a Sorting Queue Memory.

The SQM may contain several bucket queues (e.g. 4) and each SQM has the following structure:

Each physical sorting queue may contain a maximum of 256 logic queues or buckets. Each bucket is controlled by its own readpointer (*rptr*), writepointer (*wptr*), depth (*dpth*) and warning level (*wlv*). With a global setting from the PEController the bus containing the sort data is selected (*sort_source*).

The data from that bus selects the bucket and hence its pointer set. Depending on a "get" or "put" action from the bucket the data RAM is addressed with the read or the write pointer. If read and write pointer are unequal one of them may be incremented modulo the bucket size, depending on the action: get or put. The depth counter is always incremented or decremented, depending on a get or put action, without the limitation of the read and write pointer being equal. In this case the depth counters can be used as event counters and the SQM functions as a histogram module. The number of counted events can thus be larger than the bucket depth, e.g. using 18 bit depth count. A warning level can be set for each bucket. The status signal can be used in the PE-c for exception handling. All pointers, depths and warning levels can be read and written through the PE busses. The SQM is assumed to take 2 clock-cycles for a get or put action and one clockcycle for manipulating the pointer set of a bucket.

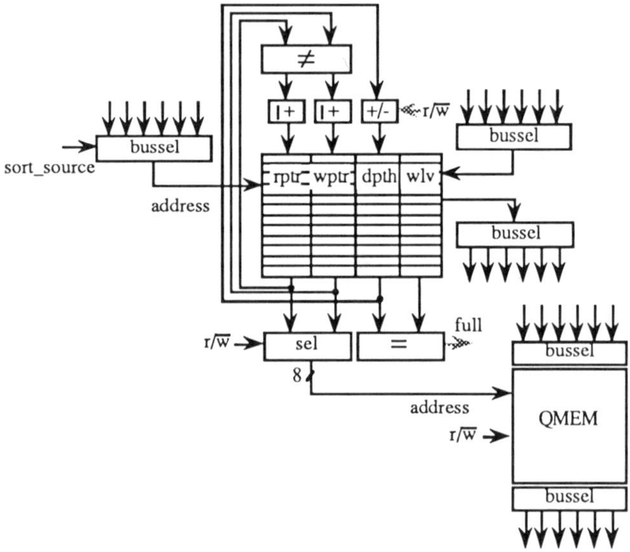

Figure 6.11 A bucket Sorting Queue Memory.

6.5 Implementation of low-level 2D Operations.

Figure 6.12a shows the way 2D images are stored in consecutive PEs for 2D Cellular Logic Operations. Each NAG of one PE can address in parallel a neighbourhood of one 2D image. Figure 6.12a shows also that the PEs, switched in a pipeline form, e.g. each processing one skeleton iteration, should have a delay between the process starting moment of the first PE and the start of the second PE. The PEs can be put into a pipeline configuration by cascading the busses, allowing each PE to write the result in memory of the next PE. As read and write cycles are excluded, contention problems are avoided.

Figures 6.12b shows the coordinated operation of the PEs for this pipeline configuration. Note, that only the used components of the datapath of the PE are drawn. The encircled numbers indicate the clock step number. The original-, mask- and result image are clocked to the next PE in each iteration.

Figures 6.12c shows that each PE is controlled by its own SSM, however, the operations of each pair of neighbouring SSMs are mutually synchronized. The figures show that after initial set-up, each pixel is processed in three clock cycles. The fact that the operation is performed in three cycles instead of two is mainly caused by the NAG-pixel counter combination, that takes 3 cycles: To clock the central pixel counter, calculate the neighbour offset and process the image edge. This addressing is pipelined with reading and writing into the image memory.

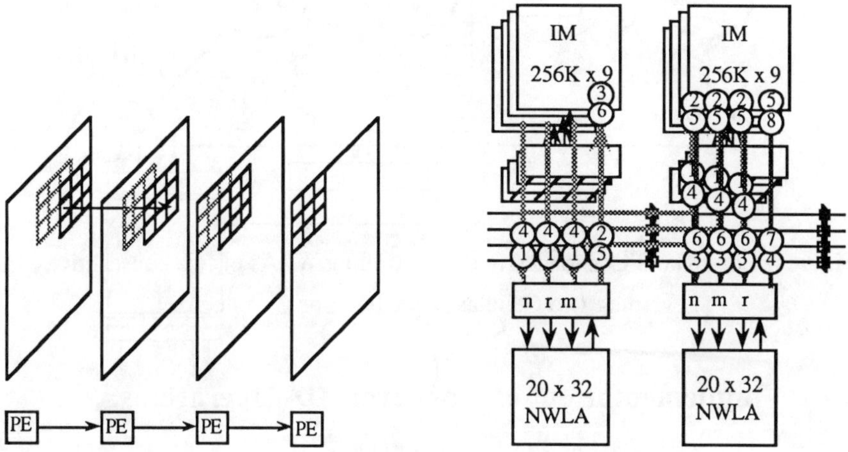

a) b)

Figure 6.12 a) Image transport and processing delay for 2D CL operations.
b) The steps for two CL iterations of one pixel.

Note that with a (very) conservative estimation of a 100n sec. clock cycle (10 Mhz) the real-time (video speed) character of the system is lost. Higher clock speeds will make real-time performance possible, as the system was functionally designed for real-time operation. The operations in this sub-section were tested on the simulated 32DIP machine.

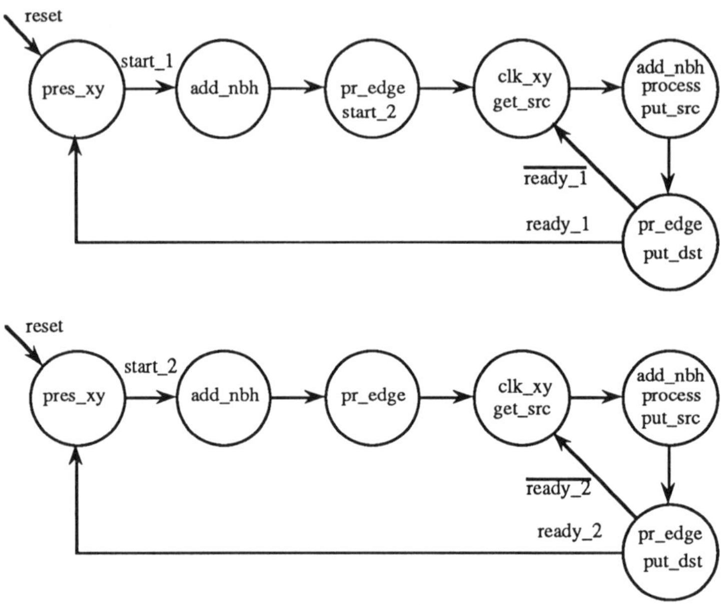

Figure 6.12c The SSMs loaded in the controlling WLAs of two consecutive PEs, governing two Cellular Logic Iterations of one pixel.

6.6 Implementation of low-level 3D Operations.

6.6.1 3D Cellular Logic Operations.

Figure 6.13a. shows how 3D binary images are processed in a 3D Cellular Logic Pipeline using RAM images instead of shift registers. Per 3D-Processing Element five images have to be maintained. The normal- and full recursive neighbourhood of slice 1, the normal- and partially recursive neighbourhood of slice 2 and the normal neighbourhood of slice 3. Optional is the mask image Z.

For 3D operations, the WLAs of a number of PEs are grouped together to form a 3D-PE. To establish the grouping, the busses of the PEs in the 3D-group are connected to each other, forming one large bus. The voxels of the addressed images are sent to the other PEs over these "segment" busses. How many PEs are grouped depends upon the necessary masks for the operation. For the 3D skeleton with endpoint condition 48 WLAs of 32 x 20 are necessary. If the WLAs in the PEs are made larger, less PEs can be used. As the busses grow larger, they need more settling time, which may cost an extra clockpulse. At this moment, however, we will pass over this problem as it depends on the actual hardware realization and wait cycles can easily be introduced.

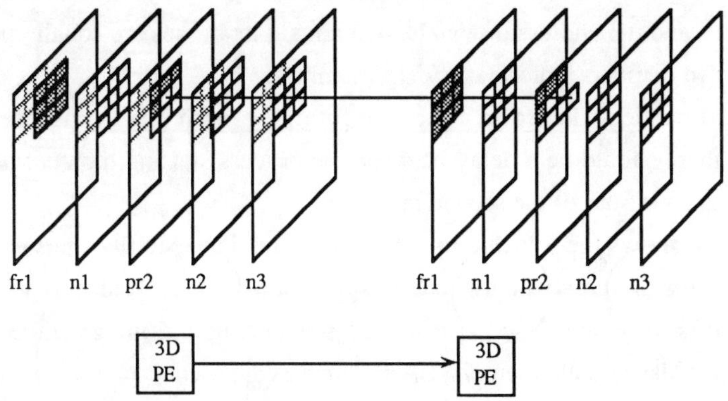

Figure 6.13a The image storage.

Figure 6.13a shows the way the 3D images are stored in a 3D-PE. Figure 6.13b shows how the 3D processing is handled by sequentially processing first the normal neighbourhood of slice 3, storing the result in the TAG register of the WLA, than processing the normal- and partially recursive neighbourhood and the mask image of slice 1 storing the result in the TAG register of the WLA and finally the normal- and fully recursive neighbourhood image of slice 1.

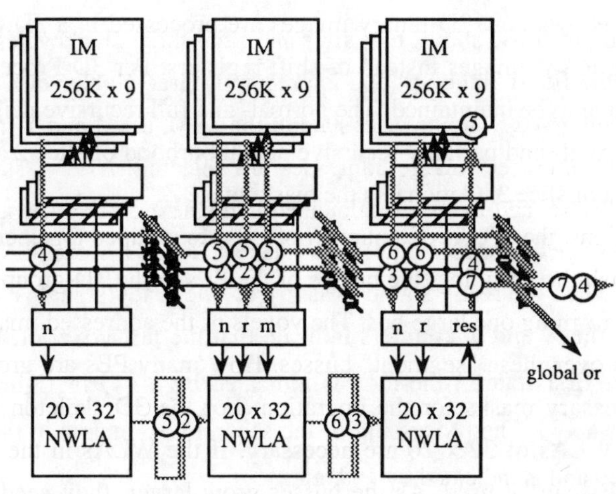

Figure 6.13b The steps for the Cellular Logic Processing of one voxel.

The WLA result is globally ORed on a bit of one of the segment busses. Beware that the three PEs in figure 6.13b indicate three steps in time and not three physically pipelined PEs. The other PEs coupled onto the busses are not drawn. The datapath components and the addressing via the two remaining busses are equally not drawn. The encircled numbers indicate clock step numbers.

If two 3D-PEs are switched in a pipeline form, e.g. each processing one skeleton iteration, there should be a delay between the process starting moment of the first processor and the start of the second processor.

Figure 6.13c shows the Synchronous Statemachine that specifies the steps that are undertaken for the processing of one voxel in one 3D-PE. Note, that within a 3D group of PEs only one PE-c performs the sequencing and image addressing, the other PEs are slaved; only their NWLA and images are used.

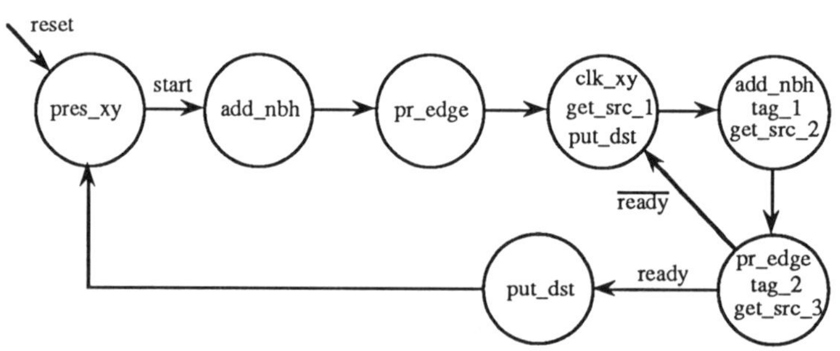

Figure 6.13c The SSM governing the Cellular Logic Processing of one voxel.

The SSM of figure 6.13c shows that after an initial three clockcycles in which the addressing of the first neighbourhood is prepared, three clockcycles are needed to produce one output voxel. In the reset state of the SSM, the X and Y image counters are preset and the correct image addresses are selected. After a start signal, the neighbourhood voxel addresses are calculated in the next clock cycle and in the third clock cycle the edge correction, if necessary, takes place. Then the SSM enters a loop that is broken when the ready signal is asserted. The ready signal is assumed to be asserted when the X and Y counters indicate that the last voxel of the image was processed. The first state in the loop is indicated by a ① in figure 6.13b. The second state by a ②. The processing of the second voxel starts in the second pass through the loop and is indicated by ④, etc.

The operations in this sub-section are tested on the simulated 32DIP machine.

6.6.2 3D Grey value operations.

For 3D greyvalue operations, both the LPA and the PL paradigm can be used. However, when data dependent processing is required the LPA paradigm can be applied best. Figure 6.14a shows a 3 x 3 x 3 neighbourhood operation and a 4 x 4 x 4 neighbourhood operation. Note, that in both cases the neighbourhoods are sparsely filled. Note also that each PE processes its own 3D slice but needs partial results from its neighbours. Suppose the sum of all neighbours in a 3 x 3 x 3 neighbourhood has to be calculated, then each PE calculates the central partial sum of {C, N, S, E, W} and then the partial sum for the PE L(eft) and the partial sum for the PE R(ight), after which all PEs add the left partial sum and then the right partial sum to their own central partial sum. If larger neighbourhoods than 3 x 3 x 3 are used, the addition of partial neighbour sums must be repeated.

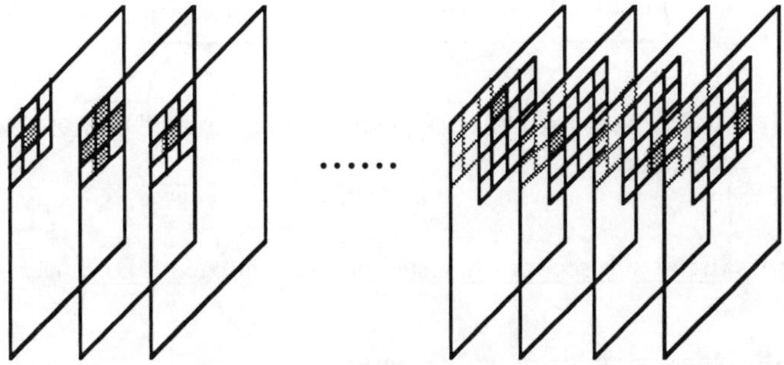

Figure 6.14a The 3D image storage of two operations in LPA mode.

Figure 6.14b shows the data transfers and the clock cycles that are necessary to process the 3 x 3 x 3 neighbourhood of figure 6.14a for three PEs switched in a LPA configuration. Note that processing is performed with a Look-up Table (LUT) and that the data selector of the LUT is provided with an accumulator that can be used for cumulative operations, without accessing the image memory. Nine clock pulses are necessary to perform the operation.

Greyvalue operations on 2D images can be performed in various ways. However, for larger neighbourhoods, the NAG is capable of addressing in its own Image Memories larger neighbourhoods sparsely, the best storage method is to store an entire 2D image in a single PE and use the PEs in a pipelined way. As pixels retrieved from a larger neighbourhoods stored in LPA mode have to pass through the busses even possibly via other PEs, the limitation on the number of busses might form a bottleneck, although the transfer can always be performed sequentially. The ring network can also be used but is relatively slow for this purpose.

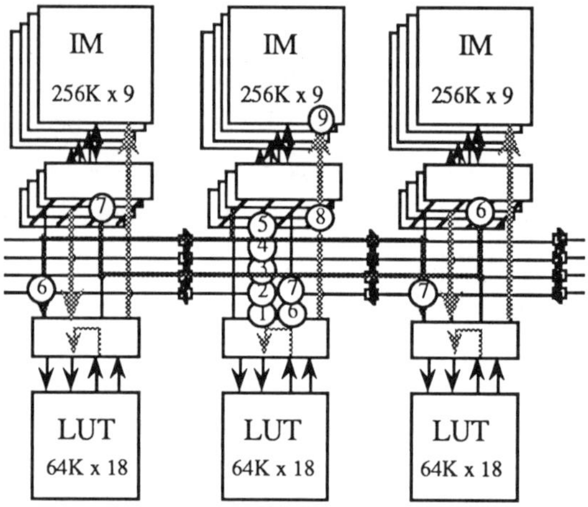

Figure 6.14b The steps for the greyvalue processing of one voxel in LPA mode.

The operations in this sub-section are tested on the simulated 32DIP machine.

6.6.3 3D Recursive grey value operations.

For 3D recursive greyvalue processing, as in the Distance Transform the Pipeline paradigm should be used. Figure 6.15a shows two 3D recursive neighbourhoods that are used in the 3D distance transform. (For the ease of explanation uniform sampling in X, Y and Z is assumed).

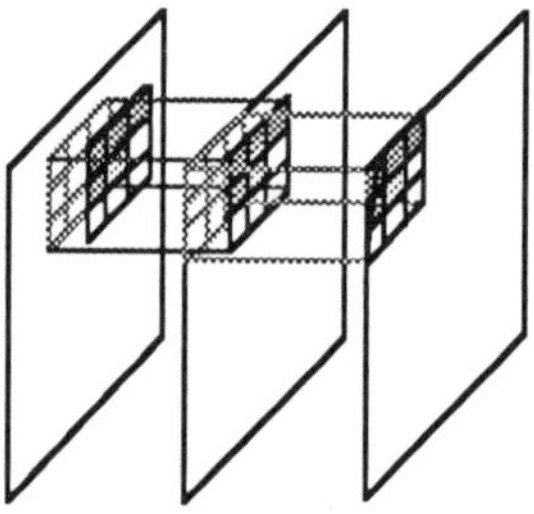

Figure 6.15a The processing of a RNO in a Pipelined LPA mode.

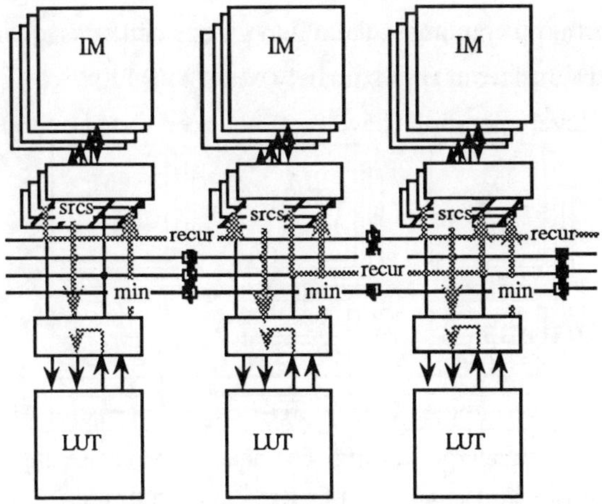

Figure 6.15b Data transfers for greyvalue RNOs in a Pipelined LPA.

Figure 6.15b shows the data transfers that are involved between three successive PEs in the Pipelined LPA.

A mixed Pipeline and Linear Processor Array paradigm can be used as was proposed in (Dekker 1987) and (Jonker 1990) for the Distance Transform on dataflow processors, but now adapted to 3D processing on the 23DIP. Successive lines of the 3D image can be processed by successive PEs, each a suitable number of pixels delayed to allow the use of recursive information from the former PE in the Pipeline as figure 6.15b shows for three PEs.
The operations in this sub-section are tested on the simulated 32DIP machine.

6.7 Intermediate level algorithms.

Low level algorithms in 2D and 3D, based on LPA or PL mode neighbourhood operations both in normal and recursive form, can be easily timed by counting clock-cycles. The described intermediate level algorithms of section 6.2 cannot be easily timed as the transfer over the rings, the FIFO queues and the data dependency of the algorithms make clock-pulse counting too complex. These algorithms have to be simulated on the architectural simulator, using various well chosen test situations. This has not yet been performed. For an estimation of the performance of the A* algorithm on the 23DIP architecture we refer to section 6.2.1. Note that PEs can be made more powerful by combining them into groups and consequently the PE of the PFM can be constructed from approximately two 23DIP PEs.

The intermediate level operations have (yet) not been tested on the simulated 32DIP machine.

6.8 Conclusions.

It is feasible to build a combined 2D and 3D-Image Processor (23DIP) for low-level and intermediate level image processing. The data rates in estimated clock cycles for a number of operations is listed in table 6.3. The Cellular Logic Operations are assumed to take place in a 3 x 3 neighbourhood in 2D and in a 3 x 3 x 3 neighbourhood in 3D. The greyvalue operations are assumed to have a kernels filled with 4 points in 2D and 6 points in 3D. Note that due to the NAGs, the actual kernel size does not matter. Also due to the NAGs, each operation on a whole image needs three additional clockcycles for the NAGs to start up. For greyvalue operations, where possible, use was made of kernel separation and storage of partial sums. The table lists the number of estimated clock-cycles per pixel in 2D and per voxel in 3D.

The (proven; measured) benefit of the proposed architecture for various intermediate level algorithms is still subject of research.

As a result of the simulations the following figures are found to be feasible for some low-level operations:

Low Level Operation types:	Single Bit nbh = 3 x 3 (x3)	Estimated Clock Cycles	Word: nbh = 4 points (2D), 6 points (3D)	Estimated Clock Cycles
Monadic Point	Not	2D: 3 3D: 3	Abs, Bit Shift, Incr, Decr, Sqrt, Sine,Ln,..	2D: 3 3D: 3
Dyadic Point	And, Or, Xor,...	2D: 3 3D: 3	Add, Sub, Threshold Mul-, Div-, Mod-pass, Max, Min	2D: 4 3D: 4
Monadic Neighbourhood Morphological	Ero, Dila, Contour,... Image Shift.	2D: 3 3D: 5	Gradient filters	2D: 4 3D: 4
Monadic Neighbourhood Statistical	Binary Rank (= Majority vote)	2D: 3 3D: 5		

Monadic Recursive Neighbourhood Morphological	Skeleton Pass	2D: 3 3D: 5		
Monadic Recursive Neighbourhood Statistical	(Recursive Binary Rank Filter pass.)	2D: 3 3D: 5		
Dyadic Point -Recursive Neighb. Morphological	Propagation pass, Anchored skeleton pass	2D: 3 3D: 5		

Table 6.3 Timing of some operations on the simulated 23DIP.

The low-level operations as tabulated in table 6.4 as well as the A* and region growing algorithms are still subject of investigation.

Low Level Operation types:	Word: nbh = 4 points (2D), 6 points (3D)
Monadic Neighbourhood Statistical	General Convolution, Gauss, LMax, LMin, Rank Filtering.
Monadic Recursive Neighbourhood Morphological	Sobel, Laplace
Monadic Recursive Neighbourhood Statistical	Fast Uniform, Laplace, Recursive Rank Filter pass.
Dyadic Point - Recursive Neighb. Morphological	Grey value propagation pass.
Dyadic Point - Recursive Neighb. Statistical	Distance Transform pass
Monadic Image Global	Pixel Count, FFT, Affine transform
Dyadic Point - Image Global	Histogram, Eql, Cst

Table 6.4. Low-level operations under investigation.

7 Conclusions.

In the previous chapters we discussed the design of high speed special computer architectures for low- and intermediate level image processing and their implementation in VLSI techniques.

Based on the experiences of past designs and the outcome of recent studies in comparisons of low-level image processing architectures, a pipelined system for real-time low-image processing was designed and realized in CMOS technology.
To overcome design pitfalls, a study was performed to the solution details found in embodiments of the three main architectural groups of image processing, the Square Processor Arrays, the Linear Processor Arrays and the Pipelines. This was reflected in a theoretical model.
As the design was based on the binary processing of images, before the actual system design, however, research had to be performed on the principles of Cellular Logic Processing. The developed methodology, based on transforming images using sets of Hit-or-Miss masks, appeared to be extendable to higher dimensional images. This resulted in a theoretical model for the generation of break-point conditions in high dimensional images, yet applied till dimension three.
Based on this result and on the requirements that intermediate level tasks pose to a system, an architecture was proposed and elaborated for the processing of low- and intermediate-level images in two and three dimensions.

In *chapter 2* the requirements for a novel architecture were condensed by reviewing current image processing tasks. These tasks lead both to low- and intermediate level operations. The requirements that execution of these operations pose on a computer system were formulated. The main conclusions were that: most image processing sub-tasks demand a mixture of low-level and intermediate level image processing operations and sole low-level tasks are currently seldom. As a consequence, novel computer architectures should smoothly support both low-level as well as intermediate-level image processing. Still, fast random access to each single pixel and the possibility to set up any data-structure is essential.

In *chapter 3* computer architectures of various natures were reviewed with the aim to discuss their competence for image processing. The conclusions were that the

memory management systems of current uni/multiprocessor systems are not optimally tuned for image processing and that this cannot be solved by using vector processors.
Special purpose systems to boost the performance of a uni/multiprocessor system must be explicitly designed for image processing, properly solving the few peculiarities of this field, such as edge handling. The same applies if real-time performance is demanded.
Most low-level image processing systems are bound too much at the image and filter-kernel data structure to allow flexible introduction of other data structures, necessary for intermediate level image processing. Pure MIMD systems are not suitable for image processing and coupled MIMD-SIMD systems must be designed such that, no bottlenecks are left, as the finest detail of the datastructures for intermediate level algorithms are usually still pixels.
Virtual systems with a single clear paradigm well supported by system software, are attractive. The attractiveness stems largely from the fact that the system leaves the programmer free within a paradigm and tries to predict the behavior of the program, resulting in the optimizing of its own configuration (by caching), rather than leaving the programmer optimize his program into a fixed datastructure. Virtual systems are certainly necessary when the number of image elements is too large for the number of available processors, as in 3D processing.

Chapter 4 was devoted to cellular logic processing and conditions for thinning in multi- dimensional images. A theoretical framework was set-up to generate sets of masks that are able to process binary images, yet applied up to dimension three.

In the first sub-section of the chapter an informal and practical approach to Cellular Logic Processing in two dimensional images was presented. The method was based on the mathematical morphology, but visualized the morphological properties of the transformation. This set-up made it possible to make variations on known transformations, to compare the effect of known transformations of different authors with eachother, and to propose new transformations.
In the second sub-section a general method has been set up for generating structuring element sets for thinning in N dimensional images.
The connectivity between image elements was defined and its relation was showed with the common indication of connectivity: The number of elements within a hypersphere in a square neighbourhood around the image element.

Objects in images were defined as compound structures composed of basic objects, whereas each basic object was defined as a non forking group of image elements with an intrinsic dimension. Object primitives were then defined as the smallest object still having the property of a basic object.
The connectivities of foreground and background objects were discussed, resulting in a scheme in which the lowest connectivity can be best chosen for the background and for the lowest to the highest intrinsic object dimension a connectivity rating from the lowest to the highest is preferably chosen.
It was showed that mask sets could be established by intersecting proper foreground and background masks. Thereafter, thinning sets with examples were presented and for sequential update techniques the role of the recursive neighbourhood was elucidated. It was shown that the mask sets could be minimized using logic minimization and that common logic operations make it possible to manipulate thinning sets to compare them. A linkage with near Euclidian skeletons was indicated.
A conclusion was, that by using this method, the original complexity of the 2^{27} possibilities in a 3 x 3 x 3 neighbourhood for three dimensional images, was brought back to a surprisingly low number of only 10 different surface masks and 34 space curve masks. The reduction from rough number of possibilities to the numbers that were found is largely on the account of the method of structuring the objects in the image. A further reduction was obtained by logic minimization.

In *chapter 5* the design of an architecture for real-time low level image processing was discussed. The design served many goals: A basis for the comparison of massive data parallel SIMD architectures and instruction parallel pipelines. Experience in VLSI design as an inevitable vehicle for future architectures. Research on mathematical morphology in multi-dimensional images, and finally, test object for VLSI design tools under development.
In the first section, the designs of two special Logic Units for cellular logic processing were discussed. A Writable Logic Array for the parallel processing of sets of masks for the class of morphological operations, and a Tally circuit for the processing of the class of binary rank filters was presented.
In the second section, the design aspects of real-time low-level image processing architectures were discussed from the designers point of view. Bottlenecks and design pitfalls on the road to a real-time system design were signalled. A theoretical model was set-up as an inventory tool for possible delays in a system.

In the second section, a design method for special computer architectures was presented. As control flow structuring may play an important role in the design of image processing architectures, the method focuses on asynchronous state machines to model this control.
In the fourth section, the architecture of a special VLSI circuit for Cellular Logic Processing was discussed. This Cellular Logic Processing Element (CLPE) is a device that is able both to operate as a single processor and as a Processing Element of a real-time binary image processing pipeline.
In last section the possibility was discussed to perform low-level greyvalue operations with CLPEs. A number of CLPEs was stacked to form a Grey Value Slice and a number of Grey Value Slices was cascaded to form a programmable pipeline for real-time low level image processing.

In *chapter 6* a system design was proposed and analyzed for a combined 2D and 3D-Image Processor for low-level and intermediate level image processing. The system was based on a combined Linear Processor and Pipeline model for low-level image processing. The system is capable to retrieve neighbourhoods in parallel for cellular logic operations as well as to retrieve large scarce neighbourhoods for greyvalue processing. Cellular logic processing was performed using a Writable Logic Array, yet for 3D Cellular Logic Operations a number of PEs need to be combined. As a service to intermediate level algorithms, bucket sorting queues were introduced as well as a mail network. The PEs were coupled onto an MIMD system, that among others acted as a long term controller, whereas a separate downloadable low-level controller was installed for the control of autonomous sub-cycles.
The benefit of the proposed architecture for various low-level algorithms was theoretically investigated and timed, while the benefit for various intermediate level algorithms was found to be subject for further research.

Some final conclusions in the field of special architectures, VLSI design and cellular logic image processing can be drawn at this point :

The perspective of future architectures for image processing is somewhat sobering. After the clear paradigms of the past decade, the Square Processor Arrays, Pyramid Architectures, Linear Processor Arrays and Pipelines, the fast workstations still seem to have the initiative. Undoubtedly the extensive and versatile capabilities of these

systems are attractive to the users, yet in many image processing applications they still fall short, due to their limited low-level image processing speed. However, they still appeal as they easily allow one to express and exploit ideas in image programming, especially when equipped with a good image processing software package.
But in the end, the slow low-level part has to be speeded up with special hardware, that is usually difficult to program.
The challenge now is to build systems that allow one to program seamlessly both in image/kernel data structures as well as in any other data structure used in intermediate level processing, yet maintaining speed at the low-level. As image data becomes larger and the number and power of PEs cannot keep up with the demand due to technical bottlenecks in interconnections, solutions must somehow be found in the creation of virtual systems, that trace and adapt their processing structures to the needs of the user, rather than proscribing standard solutions.

The conclusions within the field of VLSI design is, as with all new fields, that this area was more promising than it appeared to be. The development times are too long, and the development tools are too meager. Although this is improving steadily, the best way to use special VLSI circuits still seems to be the implementation of very regular structures, rather than building complex processors. And that is regrettable.

References

Agnew D. (1991), VHDL Extensions Needed for Synthesis and Design. In: Borrione D and Waxman R (Eds), Proceedings of the IFIP WG 10.2 Tenth International Symposium on Computer hardware description languages and their Applications. Marseille, France. North Holland. ISBN 0-444-89208-7. pp. 335-349.

Aho A., Hopcroft J., Ullman J. (1974), The design and analysis of computer algorithms, Addison Wesley.

Arcelli C., Cordella L., Levialdi S. (1975) Parallel thinning of binary pictures. Electronic letters 11: pp 140-149.

Ashcroft N.W., Mermin N.D. (1976), Solid state physics, Holt, Rinehart & Winston, New York. ISBN 0-03-083993-9.

Åström A. (1990) A smart image sensor description and evaluation of PASIC. PhD thesis (No. 257) Departement of Electrical Engineering, Linköping University, Sweden.

Barbacci M.R. (1987) An introduction to ISPS. Technical report. Dept. of computer science, Carnegie Mellon University.

Bart M., Buurman J., Duin R.P.W. (1990) A learning procedure for the recognition of 3D objects from 2D images. In Proceedings SPIE Conference Machine Vision Systems Integration in Industry, Boston Massachusetts, USA, 8-9 November.

Basille J.L., Castan S., Latil J.Y. (1981) Systeme Multiprocesseur Adapte ay Traitement d'Images. In: Duff M.J.B., Levialdi S., Languages and Architectures for Image Processing, Academic Press, London, pp. 205-213

Basille J-L., Dalle P., Castan S. (1986), Iconic and symbolic use of a line processor in multilevel structures, in: M.J.B. Duff (ed.), Intermediate level Image Processing, Academic Press, London, pp. 231-241.

Batcher K.E. (1980) Design of a massively parallel processor. IEEE Trans. Comput. C-29:836-840

Besl P.J., Jain R.C. (1986) Invariant surface characteristics for 3D object recognition in range images, Computer Vision, Graphics and Image Processing 33:33-80.

Besl P.J., Jain R.C. (1988) Segmentation through variable-order surface fitting. IEEE transactions on PAMI Vol 10, 2:167-192

Beun M. (1973) A flexible method for automatic reading of handwritten numerals. Philips Techn. Rev 31-89-101.

Blaauw G.A. (1976) Digital system implementation. Prentice Hall, Inc. Englewoods Cliffs, N.J.

Blackman T., Fox J., Rosebrugh C. (1985) The SILC siilicon compiler language and feautures. Proc. of the 22nd ACM/IEEE Design Automation Conference, Las Vegas, Nevada.

Boekamp R., Groen F.C.A., Gerritsen F.A., van Munster R.J. (1985), Design and implementation of a cellular-logic VME processor module. In: M.J.B. Duff, H.J. Siegel, F.J. Corbett (eds), Architectures and algorithms for digital image processing, Cannes, France, Proceedings SPIE Volume 596. Bellingham, Washington, SPIE, 1986, pp. 41-45.

Boomgaard R. van den (1992), Mathematical Morphology; Extensions Towards Computer Vision. PhD. Thesis. University of Amsterdam, Faculty of Mathematics and Computer Science. Amsterdam, The Netherlands.

Borgefors G. (1986) Distance transformations in digital images. Computer vision, Graphics and Image Processing 34 344-371.

Brayton R.K., Hachtel G.D., McMullen C.T., Sangiovanni-Vincetelli A.L. (1984a) ESPRESSO II A new logic Minimizer for Programmable Logic Arrays. Proceedings of the IEEE custom integrated circuits conference pp370-376.

Brayton R.K., Hachtel G.D., McMullen C.T., Sangiovanni-Vincetelli A.L. (1984b) Logic minimization algorithms for VLSI synthesis. Kluwer, Dordrecht / Boston

Brayton R.K. (1987) Algorithms for Multi-level logic Synthesis and Optimization In: De Micheli G., Sangiovanni-Vincetelli A.L., Antognetti P. (eds): Design systems for VLSI circuits; Logic Synthesis and Silicon Compilation. NATO ASI series E: Applied Sciences, No. 136, Martinus Nijhoff Publishers, Dordrecht / Boston.

Brenner (1984) The Yorktown Logic Language: An APL like design language for VLSI Specification ICCD 84.

Brown D.W. (1981) A state machine synthesizer SMS. Proceedings of the18th Design Automation Conference. Nashville. pp 301-305.

Burt P.J. (1984) The Pyramid as a Structure for Efficient Computation. In: A. Rosenfeld (Ed) Multiresolution Image Processing and Analysis. Springer Verlag Berlin.

Burt P.J., van der Wal G.S. (1987) Iconoc image analysis with the Pyramid Vision Machine (PVM). Proceedings of the 1987 workshop on computer architecture for pattern Analysis and Machine Intelligence, Seattle WA.

Burt P.J., van der Wal G.S. (1990) An architecture for multiresolution focal image analysis. Proceedings of the 10 Int. Conf on Pattern Recognition. Atlantic City. NJ.

Burt P.J., (1988) Smart Sensing in Machine Vision. In: Freeman H. (ed.) Machine vision: algorithms, architectures and systems. Academic press.

Buurman J., Duin R.P.W. (1988) Implementation and use of software scanning on a small CLIP4 processor array. In: Kittler J (ed) Lecture notes in computer science 301: Pattern Recognition. Springer Verlag, Berlin pp. 269-277.

Buurman J, Duin R.P.W. (1989) Object recognition using inexact matching of 3d graphs, Proceedings of the 5th International Conference on Image Analysis and Processing, Positano Italy, September 20-22.

Caldwell S.H (1958) Switching circuits and Logical Design. New York: Wiley & Sons, Inc.

Cantoni V. (1986) Pyramidal systems for computer vision. Nato ASI series F, vol 25. Springer Verlag, Berlin. ISBN 3-540-17165-7.

Cantoni V., Levialdi S. (1987) PAPIA: A case history. In: Uhr L (ed.) Parallel computer vision. Academic press, 3-13

Cheng X.S. (1990) Design and implementation of an image understanding system DADS. PhD thesis, TU-Delft, Information Theory Group, Faculty of Electrical Engineering.

Cooley J.W., Tukey J.W. (1965), An algorithm for the machine calculation of complex fourier series, Mathematics of Computation, Vol 19, 4:297-301

Cremers A.B., Hibbard T.N. (1985) Executable Specification of Concurrent Algorithms in the Terms of Aplicative Data Space Notation. In: Kung S.Y., Whitehouse H.J., Kailath T. (Eds). VLSI and Modern Signal Processing. Prentice Hall Information and System Sciences Series. Englewood Cliffs, New Jersey.

Danielson P.E. (1980) Euclidean Distance Mapping. Computer Graphics and Image Processing 14:227-248.

Danielsson P.E., Levialdi S. (1981) Computer architectures for pictorial information systems. IEEE Computer 14(11):53-67

Danielson P.E. (1990) Generalized and separable Sobel operators. In: Freeman H. (ed) Machine Vision. Acquiring and interpreting the 3D scene. Academic Press.

Davis R., Thomas D. (1984) Systolic array chip matches the pace of high speed processing. Electronic design, pp. 207-218.

Dekker S.T. (1987), Toward a hardware architecture for robot path finding, Msc. thesis, Pattern Recognition Group, Faculty of Applied Physics, Delft University of Technology, The Netherlands.

Dekker S.T., Jonker P.P., Groen F.C.A. (1987) Distance transform with dataflow techniques. Proceedings of the 6th Aachen symposium on signal theory, Aken, Duitsland, September 9-12, 1987. Berlin, Springer Verlag, pp. 269-272.

Dewilde P.M., Deprettere E., Nouta R. (1985a) Parallel and Pipeline VLSI Implementation of Signal processing Algorithms. In: Kung S.Y., Whitehouse H.J., Kailath T. (Eds). VLSI and Modern Signal Processing. Prentice Hall Information and System Sciences Series. Englewood Cliffs, New Jersey.

Dewilde P.M., Genderen A.J. van, de Graaf A.C. (1985b) Switch level Timing Simulation, Proceedings of the ICCAD

Duff M.J.B. (1982) CLIP4. In: Fu, K.S. and Tadao Ichikawa (eds.), Special Computer Architectures for Pattern Processing, CRC Press, Boca Raton, Florida, USA, pp. 65-86.

Duff M.J.B. (1986a) (Ed.) Intermediate Level image processing. Academic Press, London.

Duff M.J.B. (1986b) Complexity. In: Duff M.J.B. (ed) Intermediate Level image processing. Academic Press, London, pp307-314

Duin R.P.W., Jonker P.P. (1986a) Processor arrays versus pipelines for cellular logic image operations. In: Young I.T. et al.(eds) Signal Processing III: Theories and Applications, Elsevier Science Publishers, pp 1339-1342

Duin R.P.W., Haringa H., Zeelen R. (1986b) A hardware design for fast 2D percentile filtering. In: Duff M.B.J. et al. (Eds) Architectures and Algorithms for digital image processing. Proceedings of the SPIE Volume 596. Bellingham, Washington.

Duin R.P.W., Haringa H., Zeelen R. (1986c) Fast Percentile filtering Pattern recognition Letters 4. 269-272.

Duin R.P.W., Jonker P.P. (1988) Processor arrays compared to pipelines for cellular image operations. In: Uhr L. (ed) Multicomputer Vision. Academic Press, pp151-169

Duin R.P.W., Komen E.R. (1990), Massively parallel architectures for cellular logic image processing, in: V. Cantoni, L.P. Cordella, S. Levialdi and G. Sanniti di Baja (eds.), Progess in Image Analysis and Processing, World Scientific, Singapore, pp. 643-657

Eichelberger E.B., Williams T.W. (1978) A logic design structure for LSI testing. Proc. 14th Design Automation Conference. June,pp. 462-468.

Eijk P. van, Vissers C.A., Diaz M. (1989) The Formal Specification Language LOTOS: result of the ESPRIT / SEBOS project. North Holland.

Flynn M.J. (1972) Some computer organizations and their effectiveness. IEEE transations on computers C-21(9):948-960

Flynn P.J., Jain A.K. (1991) CAD-Based computer vision: From CAD models to relational graphs. IEEE Transactions on PAMI vol 13, 2:114-132

Fountain T.J. (1985), A review of SIMD Architectures, in: J. Kittler and M.J.B. Duff (eds.), Image Processing System Architectures, Research Study Press, Letchworth, UK, pp. 3-22.

Fountain T.J. (1986) Array architectures for Iconic and Symbolic image processing, ICPR-8, Paris, pp. 24-33

Fountain T.J. (1987) Processor Arrays. Academic Press, London, ISBN 0-12-262945-0

Fountain T.J., Postranecky H., Shaw G.K. (1987), The CLIP4S System, Pattern Recognition Letters, vol. 5, no. 1, pp. 41-47.

Fountain T.J. (1988a) Introducing local autonomy to processor arrays. In: Freeman H. (ed.) Machine vision: algorithms, architectures and systems. Academic press, pp.31-56

Fountain T.J., Matthew K.N., Duff M.J.B. (1988b) The CLIP7A image processor. IEEE transactions on pattern analysis and machine intelligence PAMI 10(3):310-319

Fujita Y., Iwashita M., Temma T. (1990) A dataflow image processing system TIP-4. In: Cantoni V., Cordella L.P., Levialdi S., Sanniti di Baja G. (eds) Progress in image analysis and process-ing. World Scientific, London, pp.734-741

Gerbrands J.J. (1988) Segmentation of Noisy Images. PhD. Thesis. Faculty of Electrical Engineering. Delft University of Technology, The Netherlands.

Gerritsen F.A., Aardema L.A. (1981) Design and use of the DIP-1: A Fast, Flexible and Dynamically Microprogrammable Pipelined Image Processor. Pattern Recognition, 14(1-6):319-330

Gerritsen F.A. (1982) Design and Implementation of the Delft Image Processor DIP-1. PhD thesis. Faculty of Applied Physics. Delft University of Technology, The Netherlands.

Gerritsen F.A. (1983), A Comparison of the CLIP4, DAP and MPP Processor-Array Implementations, in: M.J.B. Duff (ed.), Computing Structures for Image Processing, Academic Press, London, pp. 15-30.

Giardina, C.R. (1988) Morphological methods in image and signal processing. Prentice Hall, Englewood Cliffs-NJ. ISBN 0-13601295-7.

Golay M.J.E. (1969) Hexagonal Parallel pattern transformations. IEEE transactions on computers C-18, 8:733-740

Goldberg A., Robson D. (1983) Smalltalk-80. The Language and its Implementation. Menlo Park: Addison Wesley.

Goor A.J. van de (1989) Computer Architecture and Design. Addison Wesley Publishing Company. ISBN 0-201-18241-6.

Graaf A.J. de, Leuken T.G.R. van (1984) APS: The Simulator; APS: The Language. Technical Reports 84-10 & 84-11. Network Theory group, Faculty of Electrical Engineering, Delft University of Technology, Delft, The Netherlands.

Granlund G.H., Arvidsson J.B. (1985) Computer architectures for image processing. Proceedings of the 4th Scandinavian Conference on image analysis. Trondheim, Norway, June 17-20.

Groen F.C.A., Sanderson A.C., Schlag J.F. (1985) Symbol recognition in electrical diagrams using probabilistic graph matching. Pattern Recognition Letters, Vol 3, 5:343-350

Groen F.C.A., Jonker P.P., Duin R.P.W. (1988) Hardware versus software implementations of fast image processing algorithms. In: Jain A.K. (ed) NATO ASI series Vol F, Real-time Object Measurement and Classification. Springer Verlag Berlin.

Gurd J.R., Kirkham C.C., Watson I. (1985), The manchester prototype dataflow computer, Communications of the ACM, vol 28-1, Januari, pp34-52.

Guttag J. (1977) Abstract data Types and the Development of Data Structures. Communications of the ACM. Volume 20 No.6

Hennesy J.L. and Patterson D.A. (1990) Computer Architecture, A quantitative Approach. Morgan Kaufmann Publ. San Mateo, CA.

Hilditch C.J. (1969) Linear Skeletons from square cupboards. In: Meltzer B., Mitchie D. (Eds.). Machine Intelligence Vol. 4. (1969) University Press Edingburgh, 404-420.

Hillis W.D. (1985) The Connection Machine. MIT Press, Cambridge MA ISBN 0-262-08157-1

Hoare C.A.R. (1974) Monitors: An operating system structuring concept. Communications of the ACM. Volume 17, No 10.

Hockney R.W., Jesshope C.R. (1981) Parallel Computers. Adam Hilger Ltd. Bristol, Uk. Isbn 0-85274-752-7

Hol O. (1987) MKWPLA: a Writable PLA generator. Internal report. Faculty of Electrical Engineering, Network Theory Section. Delft University of Technology, The Netherlands.

Hol O. (1988) Mixed level simulation: Using function blocks in SLS. Internal report, Faculty of Electrical Engineering, Delft University of Technology, 2-88.

Homewood M., May D., Shepherd D., Shepherd R., The IMS T800 Transputer, IEEE Micro, October. pp 10-26.

Houten E. (1988) UNAG: A UNiversal Address Generator. Internal Report (in Dutch), Faculty of Applied Physics. Pattern Recognition Section, Delft University of Technology.

Huang T.S. (1979) A fast Two dimensional Median Filtering Algorithm. IEEE Transactions on Acoustics, Speech and Image Processing, no 1.

Hwang K., Briggs F.A. (1985) Computer Architecture and Parallel Processing. McGraw-Hill Book Co Singapore.

IEEE (1975) IEEE-488 I ANSI-MC1.1 I (IEC-625) Standard Digital Interface for Programmable Instrumentation. IEEE New York.

IEEE (1988) IEEE Std 1076-1987 IEEE Standard VHDL Language Reference Manual. IEEE New York.

Imaging-Technology (1990) ITEX 150/151 programmers manual 47-515001-02, interpreters users manual 47-515103-01.

INMOS (1988) Occam 2 reference Manual, Prentice Hall International., New York. ISBN 0-13-629312-3.

INMOS (1989) The Transputer Databook, Transputer Applications Notebook: Architecture and Software. Inmos-SGS-Thomson.

INMOS (1991) The T9000 Transputer, Products overview, Manual. Inmos-SGS-Thomson.

Inokuchi S., Matsuda F., Sato K. (1984) Range Imaging System for 3D Object Recognition, Proceeding 7th International Conference on Pattern Recognition..

Inokuchi S., Yamamoto H., Sato K. (1986) Tuned Range Finder for High Precision 3D Data, Proceedings 8th International Conference on Pattern Recognition.

Iverson K.E. (1962) A Programming Language, Wiley, New York.

Iverson K.E. (1980) Notation as a Tool of Thought, 1979 ACM Turing Award Lecture, Communications of the ACM, august 1980, vol 23, number 8.

Iwashita M., Temma T., Mizogushi M., Matsumoto K., Shuto M., Hanaki S. (1986) A dat driven VLSI Image Processor (ImPP): A LSI version of TIP. In: L. Uhr et. al. (eds) Evaluation of Multicomputers for Image Processing. Academic Press, Inc. Boston.

Jenkins R.E., Gilbert Lee Jr D. (1987) An application specific coprocessor for High-Speed Cellular Logic Operations. IEEE Micro, pp63-70

Jonker P.P. (1979) The Structuring of Software Interfaces. Msc thesis. Twente University of Technology. Enschede, The Netherlands.

Jonker P.P., Duin R.P.W. (1985) Considerations on a VLSI architecture for Cellular Logic Operations. Proceedings of the IEEE Computer Society workshop on Computer Architecture for Pattern Analysis and Image Database Management, Miami Beach, FL: 453-462

Jonker P.P. (1985) Internal report MP 940. TNO-Institute of Applied Physics, Delft, The Netherlands. 10-11-85.

Jonker P.P., Dekker S.T., Verwer B.J.H. (1988a), A hardware architecture for robotpath planning, Proc. of IAPR Workshop on computer vision, Special hardware and industrial applications, Tokyo, Japan, pp.100-104.

Jonker P.P., Nouta R., Kraaijveld M.A., Schot C.A. (1988b) The realization of a VLSI Circuit for fast binary image processing using a new VLSI design system. In: Herrmann O.E., Beijnum B.J.F. van (eds) Lecture Notes of the Nelsis-Project. TU-Twente, pp62-83.

Jonker P.P. (1988c) Editor, The Architecture of the Delft Intelligent Assembly Cell (DIAC), TU-Delft, December 1988, Milestone 1 report SPIN/FLAIR-2 project IV-1.1

Jonker P.P., Komen E.R., Duin R.P.W., Kraaijveld M.A. (1989) A Cellular Logic Pipeline for Real Time Robot Vision, Internal Report TUD/PH 89.02 Faculty of Applied Physics. Delft University of Technology, The Netherlands.

Jonker P.P., Zeppenfeldt F., Dekker S.T., Duin R.P.W. (1990a), Image processing using data flow based digital signal processors, Proceedings of the Workshop on Parallel Processing, BARC Bombay 400 085, India, pp. IP14-IP26.

Jonker P.P., Komen E.R., Duin R.P.W. (1990b), Architectures for 2D to 3D Low Level Image Processing, IAPR Workshop on Machine Vision and Applications, Tokyo

Jonker P.P., Schmidt W.F., Verbeek P.W. (1990c) A DSP based range sensor using time sequential binary space encoding. Workshop on parallel processing BARC, Bombay, India.

Jonker P.P., Komen E.R. (1992) A scalable Real-Time Image Processing Pipeline. Proceedings of the 11th ICPR, Vol. 4: Architectures for Vision and Pattern Recognition. The Hague, The Netherlands. pp. 142-146.

Jonker P.P., Gerbrands J.J. (1992) Image Processing Hardware for Counting Massive Object Streams. Proceedings of the 11th ICPR, Vol. 4: Architectures for Vision and Pattern Recognition. The Hague, The Netherlands. pp. 31-33.

Jonker P.P., Komen E.R., Kraaijveld M.A. (1992) A scalable Real-Time Image Processing Pipeline. Accepted for: Machine Vision and Applications. Springer.

Jonker P.P, Vossepoel, A.M. (1992) Conditions for Multidemensional Thinning; I: Analysis, II: Detection. Submitted to: IEEE journal on Pattern Analysis and Machine Intelligence.

Kanade T., Gruss A., Carley L.R. (1989) A VLSI Sensor based rangefinding system. 5th ISRR, Tokyo Japan.

Karnaugh M. (1953) The map method for synthesis of Combinatorial Logic Circuits. Transactions of the AIEE vol 72.pt I pp.593-598.

Kernighan B.W., Ritchie D.M. (1988) The C programming language (2nd edition), Prentice Hall, Englewood Cliffs, NJ 07632, USA. ISBN 0-13-110362-8.

Kimmel M.J., Jaffe R.S., Manderville J.R., Lavin M.A. (1985) MITE: Morphologic image transform engine, an architecture for reconfigurable pipelines of neighbourhood processes. IEEE Computer Society Workshop for Pattern Analysis and Image Database Management, Miami Beach, Florida:493-500

Kittel C. (1986) Introduction to solid state physics. John Wiley & Sons, Inc. New York. ISBN 0-471-87474-4

Komen E.R., Duin R.P.W. (1989) The use of Cellular Logic Processing Elements for grey-Value Pipelined Processing. Internal Report TUD/PH 89.01 Faculty of Applied Physics. Delft University of Technology, The Netherlands.

Komen E.R. (1990a), Low-level Image Processing Architectures Compared for some Non-linear Recursive Neigbourhood Operations, Phd Thesis, Faculty of Applied Physics, Delft University of Technology, Delft, 190 pages.

Komen E.R. (1990b), View Angle Transformations, Proc. of the First Int. Conf. on Information Technologies for Image Analysis and Pattern Recognition (ITIAPR '90), Lviv, USSR, pp.317-323.

Komen E.R., Teeuw W.B., Duin R.P.W., Jonker P.P. (1990c), Mapping an N*N Image onto P Processors, IAPR Workshop on Machine Vision and Applications, Tokyo, November 28-30.

Komen E.R. (1991a), Efficient parallelism using indirect addressing in SIMD processor arrays, Pattern Recognition Letters No. 12 pp. 279-289.

Komen E.R. (1991b) View Angle Transformations, Pattern Recognition Letters No. 12, pp. 273-278.

Kraaijveld M.A., Jonker P.P., Nouta R., Duin R.P.W. (1986a) The VLSI realisation of a binary-image processor, In Proceedings Signal Processing III: theories and applications. Amsterdam, North-Holland, 1986, pp. 1231-1234.

Kraaijveld M.A. (1986b) The VLSI realisation of a binary-image processor, (Msc. thesis) Report 940, Institute of Applied Physics-TNO, section MP, Delft, The Netherlands, 27-2-86.

Kraaijveld M.A. (1987) The Switch level description of the Cellular Logic Processor V2.0. Internal report, Pattern Recognition Section. Faculty of Applied Physices, Delft University of Technology, The Netherlands.

Krekelberg D.E., Sobelman G.E., Jhon C.S. (1985) Yet Another Silicon Compiler. Proc. of the 22nd ACM/IEEE Design Automation Conference, Las Vegas, Nevada.

Kung H.T. (1982) Why systolic Architectures?. IEEE Computer C-12(1)

Kung H.T. (1984) Putting Inner Loops Automatically in Silicon. In: T.L. Kunii, VLSI engineering; Beyond software engineering. Lecture Notes in Computer Science 163. Springer Verlag Berlin.

Kung H.T., Menzilcioglu O. (1984) WARP: a programmable systolic array processor. Proceedings SPIE, Vol 495, Real-time Signal Processing VII.

Lam L., Lee S.W., Suen C.Y. (1991) Thinning Methodologies - A comprehensive study. Accepted for publication in IEEE transactions on PAMI.

Lange A.A.J. de (1991) Design and Implementation of Highly Parallel Pipelined VLSI systems. PhD Thesis. Faculty of Electrical Engineering. Delft University of Technology, The Netherlands.

Lee S.W., Lam L., Suen C.Y. (1991) Performance Evaluation of Skeletonization Algorithms for Document Image Processing. Proceedings of the First Int. Conf. on Document Analysis and Recognition. ICDAR 91. Saint Malo France.

Leung S.S., Shanblatt M.A. (1989) ASIC System design with VHDL: A paradigm. Kluwer Academic Publishers, Norwell MA, USA, ISBN 0-7923-90932-6.

Levialdi S. (1988) Computer architectures for image analysis, Proc. ICPR-9, Rome, pp. 1148-1158

Lindskog B. (1988) PICAP3 An SIMD architecture for multi-dimensional signal processing. PhD thesis. Dept. of Electrical Engineering. Linköping University, Sweden.

Liskov B. (1974) Programming with abstract data types. ACM SIGPLAN Notices, Vol. 9, No 4.

Lobregt S., Verbeek P.W., Groen F.C.A. (1980) Three dimensional skeletonization: Principle and algorithm IEEE Transactions on Pattern Analysis and Machine Intelligence 2, 75-77.

Lougheed R.M. (1987) Advanced image-processing architectures for machine vision. SPIE conference on image pattern recognition. Algorithm implementations, techniques and technology: Critical review of technology. (vol. 755)

Lougheed R.M., McCubbrey D.L. (1980) The cytocomputer: a practical pipelined image processor. Proc. of 7th annual international symposium on computer architecture, pp. 271-277

Lozano-Pérez T. (1983), Spatial Planning: A Configuration Space Approach, IEEE Transactions on Computers, Vol. C-32-2.

Lozano-Pérez T. (1987), Robot programming, In: AI in the 1980's and beyond, W.E.L. Grimson and R.S. Patil, ed., MIT Press.

LSI-Logic (1989) 64xxx series Digital Signal Processing and Image Processing data sheets.

McClusky E.J. (1956) Minimization of Boolean Functions. Bell System Technical Journal vol. 35 pp 1417-1444.

McClusky E.J. (1965) Introduction to the Theory of Switching Circuits. McCraw-Hill, New York.

McCubbrey D.L., Lougheed R.M. (1985) Morphological image analysis using a raster pipeline processor. IEEE Computer Society Workshop for Pattern Analysis and Image Database Management, Miami Beach, Florida, pp 444-452

Mead C., Conway L. (1980) Introduction to VLSI systems. Addison Wesley, ISBN 0-201-04358-0.

Meijer B.R. (1987) In Dutch: PAPS een real-time beeldbewerkingssysteem (PAPS a real-time image processing system) Msc. thesis. Faculty of Applied Physics, Pattern Recognition Section. Delft University of Technology, The Netherlands.

Mokhoff N. (1985) Concurrent computers make scientific computing affordable.. Computer design, April 1985 pp. 59-60

Motorola (1988) MC88100 RISC Microprocessor Users Manual, MC88200 CMMU Users Manual. Motorola Inc.

Motorola (1990) DSP96002, 96-bit general purpose IEEE Floating point dual port processor, technical data. Motorola Inc.

Mukherjee A. (1986) Introduction to Nmos and Cmos VLSI system design, Prentice hall 1986.

Mullikin J. (1992) Personal communication.

NEC (1985) uPD7281 Users manual, NEC Electronics Corp.

Nicoud J.D., Tyrrell A.M. (1989) The Transputer T414 instruction set. IEEE Micro, June pp 60-75.

Ozaki Y, Sato K and Inokuchi S (1988) Rule-driven processing and recognition from range image. IEEE Conference CH2614-6/88/0000/0804$01.00

Parnas D.L. (1972a) A Technique for Software Module Specification with Examples. Communications of the ACM. Volume 15, No 5.

Parnas D.L. (1972b) On the Criteria To Be Used in Decomposing Systems into Modules. Communications of the ACM. Volume 15, No 12.

Pass S. (1985), The GRID Parallel Computer System, in: J. Kittler and M.J.B. Duff (eds.), Image Processing System Architectures, Research Study Press, Letchworth, UK, pp. 23-35

Pearl J. (1984) Heuristics; Intelligent Search Strategies for Computer Problem Solving. Addison Wesley Publ. Comp, Reading Massachusetts.

Persoon E. (1986) Hierarchical correlation for fast industrial object location. Pattern Recognition in Practise 2, eds. Kanals & Gelsema, North Holland 1986, p199.

Persoon E. (1988) A pipelined image analysis system using custom integrated circuits. IEEE Transactions on PAMI vol. 10 January 9188 pp. 110-116.

Peterson J.L. (1981) Petri net theory and the modelling of systems. Prentice Hall, Englewood Cliffs, New Jersey.

Plessey (1986) PDSP16401, 2-dimensional edge detector data sheet. Plessey Semiconductors. (Sobel / Prewitt edge detector chip)

Pratt W.K., Leonard P.F. (1986) Review of Machine Vision Architectures. Proceedings SPIE Conf Vol 755.

Preston K. (1991) Three dimensional mathematical morphology. Image and Vision Computing, Vol 9, No 5.

Preston K. Jr, Kaufman A.G., Sobey C. (1984) VLSI Implementation of Cellular Logic Processors. In: K.S. Fu. VLSI for Pattern Recognition and Image Processing. Springer Verlag, Berlin

Preston K. Jr. (1970) Feature extraction by golay hexagonal pattern transforms. IEEE Symposium on feature extraction and selection in pattern recognition, Argonne, III

Preston K. Jr. (1989) Benchmark Results. The Abingdon Cross . In: Uhr L. Evaluation of Multicomputers for Image Processing. Academic Press, Inc. Boston.

Preston K. Jr. (1989) The Abingdon Cross Benchmark Survey. IEEE Computer, Vol 22, No 7. pp 9-18.

Preston K. Jr., Duff M.J.B. (1984) Modern Cellular Automata, Theory and applications. Plenum Press, New York.

Quine W.V. (1952) The Problem of Symplifying Truth Functions. American Mathematical Monthly Vol. 59 :521-531.

Reeves A.P. (1984) Parallel computer architectures for image processing. Computer Vision, Graphics, and Image Processing 25:68-88

Ridler T.W., Calvard S. (1978) Picture thresholding using an interactive selection method. IEEE Transactions on Systems, Man and Cybernetics, Vol 8, 8:630-632.

Roger D., Adams F., Alan J. (1976) Mathematical Elements for Computer Graphics, McGraw-Hill, New-York.

Rosenfeld A. (1988) Hough transform algorithms for mesh-connected SIMD parallel processors. Computer vision, graphics and image processing 41:293-305

Rosenfeld A., ed (1984) Multiresolution image processing and analysis, Springer, Berlin

Rosenfeld A., Pfalts J.L. (1966) Sequential operations in Digital Image Processing. Journal of the ACM, pp471-494.

Rowen C., Hennesy J.L. (1985) SWAMI a flexible logic implementation system. Proc. of the 22nd ACM/IEEE Design Automation Conference, Las Vegas, Nevada.

Sanz J.L.C., Cypher R.E. (1988) Algorithms for massively parallel image processing architectures, ICPR-9, Rome, pp. 412-419

Schaefer D.H. (1986) Pyramid Architectures. In: Duff M.J.B. (ed.) Intermediate level image processing, Academic Press, New York, pp. 167-179

Schmidt W.F. (1989) Segmentation and Analysis of Range Images. Msc. Thesis. Pattern Recognition Section. Faculty of Applied Physics. Delft University of Technology, The Netherlands.

Schmidt W.F., Nobel J., Klop F. (1988) In Dutch: Ontwerp van een universele WLA (The design of a universal WLA). Internal report. Faculty of Electrical Engineering, Network Theory Section & Faculty of Applied Physics, Pattern Recognition Section. Delft University of Technology, The Netherlands.

Schmitt L.A., Wilson S.S. (1988) The AIS-5000 Parallel processor. IEEE Transations on pattern analysis and machine intelligence PAMI-10(3):320-330

Schot C.A. (1988) Circuit redesign and lay-out definition of the Cellular Logic Processing Element (CLPE). Msc Thesis. Faculty of Electrical Engineering, Network Theory Section Delft University of Technology, The Netherlands.

Scil_image (1991) Users manual, issue for the Centre of Image Processing and Pattern Recognition (CBP) The Netherlands. University of Amsterdam, Computer systems group, Faculty of Mathematics and Computer Science.

Serra J. (1982) Image analysis and mathematical morphology. Academic Press, Inc. London.

Serra J. (1988) Image analysis and mathematical morphology. Volume II: Theoretical Advances. Academic Press, Inc. London.

Smith W.W., Sullivan P. (oct 1984) Systolic array chip matches the pace of high speed processing, electronic Design, pp 207-209

SPARC (1990) SPARC RISC Users Guide. Ross Technology, Inc. subsidiary of the Cypress Semiconductor Company, 3901 NF-street San Jose, CA 95134.

Sternberg S.R., Cytocomputer real-time pattern recognition, Proc. 8th Auto. Imagery Pattern Recog. Symp., pp. 205-214, 1978.

Stok L. (1991) Architectural Synthesis and Optimization of Digital Systems. PhD thesis. Technical University of Eindhoven, The Netherlands.

Stonefield (1986) CLIP4 System overview, Software overview, Users guide, Programmers guide. Stonefield Ltd, Horsham UK.

Stout Q.F. (1987) Pyramid algorithms optimal for the worst case. In: Uhr L (ed) Multicomputer Vision. Academic Press, pp147-168

Stroustrup B. (1985) The C++ Programming Language. Addison Wesley Publ. Comp. Reading MA.

Stuivinga M., Nobel J., Verbeek P.W., Steenvoorden G.K. and Joon M.M. (1989), Range finding based on a position sensitive device array. Sensors and actuators 17 pp. 255-258.

Tanenbaum A.S. (1990) Structured Computer Organization. Prentice Hall. ISBN 90-6233-576-4.

Tanimoto S. (1984) Sorting, Histogramming and other Statistical Operations on a Pyramid Machine. In: A. Rosenfeld (ed) Multiresolution Image `processing and Analysis. Springer Verlag, Berlin.

Tanimoto S.,(1986) Paradigms for pyramid machine algorithms. In: Cantoni V and Levialdi S (Eds), Pyramidal Systems for Computer Vision, Berlin, Springer Verlag.

Tanimoto S. (1988) Machine vision as State Space Search. In: Freeman H (ed.) Machine vision: algorithms, architectures and systems. Academic press.

TCL-Image (1990) TCL-Image Users Manual 4.5, TCLI-90.001: Multihouse TSI b.v. & TNO-Institute of Applied Physics, Delft, The Netherlands.

Teeuw W.B. (1989) Pyramid based image processing on a CLIP4 processor array. Master Thesis Dept. of Applied Physiscs, Delft Univ. of Technology

Teeuw W.B., R.P.W. Duin (1990), An Algorithm for benchmarking a SIMD pyramid with the Abingdon Cross, Pattern Recognition Letters, vol. 11, pp. 501-506.

Thissen F.L. (1985) The Philips modular picture acquisition and processing system - PAPS. pp. 587-598. Proceedings of the 5th Intl. Conf. on Robot vision and sensory controls, 29-31 october 1985, Amsterdam.

TIM (1989) TIM users manual. DIFA Measuring Systems B.V. Druivenstraat 25 4816 KB Breda, The Netherlands.

Toriwaki J., Yokoi S., Yonekura T., Fukumura F. (1982) Topological properties and topological-preserving transformation of a three dimensional binary picture. Proc. International Conference on Pattern Recognition, Munich, pp 414-419.

Uhr L., Schmitt L. (1984) The several steps from Icon to Symbol Using Structured Cone/Pyramids. In: A. Rosenfeld. Multiresolution Image Processing and Analysis. Springer Verlag Berlin.

Uhr L. (1987) Highly Parallel, Hierarchical, Recognition Cone Perceptual Structures. In: Uhr L. (ed) Multicomputer Vision. Academic Press, pp 249-292

Van Vliet L.J., Verwer B.J.H. (1988) A contour processing method for fast binary neighbourhood operations, Pattern Recognition Letters 7, pp. 27-36

Wal G.S. van der (1991) The Sarnoff Pyramid Chip. Workshop on Computer Architecture for Machine perception CAMP-91, Paris, France, December pp 69-79.

Venema J. (1987) In Dutch: De TDCLPT1, een proefchip voor de CLPE. (The TDCLPT1, a probe chip for the CLPE) Internal Report. Faculty of Electrical Engineering, Network Theory Section. Delft University of Technology, The Netherlands.

Verbeek P.W., Dorst L., Verwer B.J.H., Groen F.C.A. (1986) Collision avoidance and path finding through constrained distance transformation in robot state space Preprints Intelligent Autonomous Systems. Amsterdam pp 643-641.

Verbeek P.W., Vrooman H.A., Vliet L.J. van (1988) Low-level image processing by max/min filters. Signal Processing 15(3):249-258

Verwer B.J.H. (1988) Improved metrics in image processing applied to the Hilditch skeleton. Proceedings International Conference on Pattern Recognition, Rome 1988, pp 137-429.

Verwer B.J.H., Dekker S.T., Jonker P.P., Groen F.C.A. (1988) Robot Path Planning by Heuristic Search. Proceedings of the 12th EMACS World Congress on Scientific Computation PARIS, july 18-22.

Verwer B.J.H. (1990) A multiresolution work space, multiresolution configuration space approach to solve the path planning problem. Proceedings. IEEE Int. Conference on Robotics and Automation, Cincinnati, Ohio, pp 2107-2112.

Verwer B.J.H. (1991) Distance Transforms; Metrics, Algorithms and Applications. Ph.D. Thesis (Ch 5.) Faculty of Applied Physics, TU-Delft, The Netherlands.

Visilog (1988) Noesis Users Guide, Programmers Guide. NOESIS, Z.A. Les Metz 5bis Rue du Petit Robinson, 78350 Jouy en Josas, France.

Vissers C.A. (1977) Interface: Definition, Design and Description of the Relation of Digital System Parts. PhD thesis. Faculty of Electrical Engeneering, Twente University of Technology, Enschede, The Netherlands.

Vuylsteke P, Oosterlinck A, (1987) 3D perception with a single binary coded illumination pattern. Optics, Illumination and Image sensing for Machine Vision. Svettkoff DJ (Ed). Proc. SPIE 728

Webb J.A., Kanade T. (1986) Vision on a Systolic Array Machine. In: L.Uhr et. al. (eds) Evaluation of Multicomputers for Image Processing. Academic Press, Inc. Boston.

Wilson SS (1988) One dimensional SIMD architectures - the AIS-5000. In: Uhr L (eds) Multicomputer Vision, Academic Press, London, pp.131-149

Yamada H. (1984) Complete euclean distance transformation by parallel operation. Proc. of the 7th Int. Conf. on Pattern Recognition: 69-71. Montreal, Canada.

Young IT (1988) Sampling density and quantitative microscopy. Analytical and quantitative cytology and histology, vol 10 pp 269-275

Young IT (1989) Image Fidelity: Characterizing the imaging transfer function, Methods in Cell Biology, Taylort DL and Wang YL (Eds), Academic Press San Diego, vol. 3B pp1-45

Zeelen R. (1986) Percentile Filter Hardware. Report 316.307. TNO Institute of Applied Physics (TPD-TNO), Delft, The Netherlands.

Zeelen R., Jonker P.P. (1987) Design sessions 2/3D UNAG. Personal Communications.

Zwart A.P.J. de (1991) In Dutch: Testen van WLA chips. Internal report ER 91.009. Faculty of Applied Physics, Pattern Recognition Section. Delft University of Technology, The Netherlands.

Samenvatting

Gebaseerd op oplossingen uit bestaande ontwerpen van de laatste jaren, alsmede op de resultaten van recente studies op het gebied van vergelijkingen van laag-nivo beelbewerkingscomputer architecturen, is er een pijplijn systeem ontworpen en in CMOS technologie gerealiseerd, voor het op video snelheid verwerken van beelden.

Om te voorkomen dat onvolkomenheden van bestaande ontwerpen wederom in een nieuw ontwerp sluipen, is er een studie verricht naar de details van de oplossingen die gevonden werden in de drie hoofd architectuur groepen van de laag-nivo beeldbewerking: De vierkante processor arrays, de lineaire processor arrays en de pijplijn computers. De uitkomst van deze studie is gecondenseerd in een theoretisch model.

Daar het ontwerp gebaseerd is op het bewerken van tweewaardige beelden, is er studie verricht naar de principes van cellulaire logische bewerkingen op twee dimensionale beelden. Een methodologie is ontwikkeld, gebaseerd op de transformatie van beelden met behulp van sets van "Hit" of "Miss" maskers. Deze methodologie bleek uitbreidbaar te zijn naar het transformeren van beelden met een hogere dimensie. Een theoretisch model voor het genereren van breekpunt condities voor hoog dimensionale beelden is ontwikkeld en toegepast en geverifieerd tot en met dimensie drie.

Gebaseerd op de resultaten van dit onderzoek alsmede gebaseerd op de vereisten die midden nivo beeldbewerkings taken stellen aan een computer systeem, is een architectuur opzet gemaakt en uitgewerkt voor het op laag- en midden nivo bewerken van twee en drie dimensionale beelden.

Appendices.

Appendix A

(Pseudo Euclidian skeletons are obtained if the candidate pixels are eroded according to their distance instead of to the erosion masks. Speed-up of a software implementation is obtained if a bucket sorting queue mechanism is used for the candidate points instead of pure raster scanning (Verwer 1988). Modest speed-up can be obtained by skipping over the 0's in the image.)

Maskset for thinning in 2D (a).

(Equivalent with the Hilditch skeleton)

Erosion mask for the erosion of 4-connected surfaces (= 8-connected contours).
(Apply in the normal neighborhood)

```
.1.
111
.1.
```

Break point mask set for the preservation of 8-connected curves in 2d.
(Apply in the recursive 0 neighborhood)

```
.1.   .0.   ..1   1..   .01   10.   ...   ...
010   111   010   .10   110   .11   011   110
.1.   .0.   1..   .01   ...   .0.   10.   .01

..1   10.   .1.   .1.   101   ...   ..1   1..
010   01.   .10   010   .10   .10   .10   010
.1.   .1.   .01   1..   ...   101   .01   1..
```

Break point mask for the preservation of a point in 2d.
(Apply in the recursive 0 neighborhood)

```
000
010
000
```

End point mask set for the preservation of 8-connected curve ends in 2d.
(apply in the normal neighborhood)

```
001   010   100   000   000   000   000   000
010   010   010   110   010   010   010   011
000   000   000   000   100   010   001   000
```

Maskset for thinning in 2D (b).

Erosion mask for the erosion of 4-connected surfaces (= 8-connected contours).
(Apply in the normal neighborhood)

```
.1.
111
.1.
```

Break point mask set for the preservation of 4-connected curves in 2d.
(Apply in the recursive 0 neighborhood)

```
.1.    .0.    .10    01.    ...    ...
010    111    .11    11.    .11    11.
.1.    .0.    ...    ...    .10    01.
```

Break point mask for the preservation of a point in 2d.
(Apply in the recursive 0 neighborhood)

```
000
010
000
```

End point mask set for the preservation of 4-connected curve ends in 2d.
(apply in the normal neighborhood)

```
010    000    000    000
010    110    010    011
000    000    010    000
```

Maskset for thinning in 3D.

(The unmirrored, unrotated sets)

Erosion mask for the erosion of 6-connected volumes in 3d.
(normal neighborhood)

```
... .1. ...
.1. 111 .1.
... .1. ...
```

Break point masks for the preservation of 18-connected surfaces in 3d.
(recursive 0 neighborhood)

```
aa           ab           ac           ad1          ad2
... ... ...  ... ... ...  ... ... ...  ... ..1 ...  ... ..1 ...
.1. 110 ...  .1. 01. ...  .1. 111 .0.  .1. 01. ...  .1. .10 .0.
... .01 .1.  ... 101 .1.  ... .0. .1.  ... 10. .1.  ... 1.. .1.

bb1          bb2          bc1          bc2
... ... .1.  ... .0. .1.  ... ... ...  ... ... ...
.0. .1. 101  ... .1. 101  .1. .10 .1.  .1. 01. .1.
... ... .1.  ... ... .1.  ... 101 ...  ... 101 ...

bd1          bd2          bd3          bd4
.1. .0. ...  .1. ... ...  .1. ... ...  .1. ... ...
... .1. ...  ... 01. ...  ... .10 ...  ... .1. .0.
... 101 .1.  ... 101 .1.  ... 101 .1.  ... 101 .1.

cc1          cd1          cd2          cd3
... .0. ...  ... ..1 ...  ... .01 ...  ... ..1 ...
.1. 111 .1.  .1. 01. .1.  .1. .10 .1.  .1. 010 .1.
... .0. ...  ... 10. ...  ... 1.. ...  ... 1.. ...

dd1          dd2          dd3
.1. .01 ...  .1. ..1 ...  .1. ..1 ...
.0. .1. ...  .0. .1. .0.  .0. 01. ...
... 1.. .1.  ... 1.. .1.  ... 1.. .1.
```

Breakpoint masks for the preservation of 26-connected curves in 3d.
(recursive 0 neighborhood)

```
1.. ... ...  .1. ... ...  ..1 ... ...  ... .1. ...  ... ..1 ...
000 010 000  000 010 000  000 010 000  000 010 000  000 010 000
1.. ... ...  1.. ... ...  1.. ... ...  1.. ... ...  1.. ... ...

... ... ..1  .1. ... ...  ... 1.. ...  ... .1. ...  ... ... .1.
000 010 000  000 010 000  000 010 000  000 010 000  000 010 000
1.. ... ...  .1. ... ...  .1. ... ...  .1. ... ...  .1. ... ...

..1 ... ...  ... ... 1..  ... .1. ...
000 010 000  000 010 000  000 010 000
..1 ... ...  ..1 ... ...  ... .1. ...
```

```
1.. 000 1..   .1. 000 1..   ..1 000 1..   ... 000 1..   ... 000 1..
... 010 000   ... 010 000   ... 010 000   1.. 010 000   .1. 010 000
... ... ...   ... ... ...   ... ... ...   ... ... ...   ... ... ...

... 000 1..   ... 000 1..   ... 000 1..   ... 000 1..   1.. 000 .1.
..1 010 000   ... 010 000   ... 010 000   ... 010 000   ... 010 000
... ... ...   1.. ... ...   .1. ... ...   ..1 ... ...   ... ... ...

.1. 000 .1.   ... 000 .1.   ... 000 .1.   ... 000 .1.   ... 000 .1.
... 010 000   1.. 010 000   .1. 010 000   ... 010 000   ... 010 000
... ... ...   ... ... ...   ... ... ...   1.. ... ...   .1. ... ...

1.. .00 .01   .1. .00 .01   ..1 .00 .01   ... .00 .01   ... .00 .01
... .10 .00   ... .10 .00   ... .10 .00   1.. .10 .00   .1. .10 .00
... ... ...   ... ... ...   ... ... ...   ... ... ...   ... ... ...

... .00 .01
... .10 .00
1.. ... ...
```

Break point mask for the preservation of a point in 3d.
(recursive 0 neighborhood)

```
000 000 000
000 010 000
000 000 000
```

End point masks for the preservation of 18-connected surfaces edges in 3d.
(normal neighborhood)

```
3a18          3a6           3b181         3b182         3b6
... .0. ...   ... .0. ...   ... .0. ...   ... .0. .1.   ... .0. ...
.0. 110 .0.   .0. 110 .1.   .0. 010 .0.   .0. 010 .0.   .0. 010 .1.
... .01 .1.   ... .01 ...   ... 101 .1.   ... 101 ...   ... 101 ...

3c18          3c6           3d18          3d6
... .0. ...   ... .0. ...   ... .01 ...   ... .01 ...
.0. 111 .0.   .0. 111 .1.   .0. 010 .0.   .0. 010 .1.
... .0. .1.   ... .0. ...   ... 10. .1.   ... 10. ...

2_18/18       2_18/6        2_6/6
... .0. ...   ... .0. ...   ... .0. ...
.0. 010 .0.   .0. 010 .1.   .0. 110 .1.
... .01 .1.   ... 10. ...   ... .0. ...
```

End point masks for the preservation of 26-connected curve points in 3d.
(normal neighborhood)

```
000 010 000   000 001 000   000 000 001
000 010 000   000 010 000   000 010 000
000 000 000   000 000 000   000 000 000
```

Appendix B.

Additional masks found in the (Arcelli 1975) skeleton compared with the (Hilditch 1969) skeleton. This makes the resulting skeleton not truly 8-connected.
Break point mask set preserving the following 4-connected situations in 2d.

```
.11   11.   .1.   .1.   .1.   .1.   10.   .01   01.   .10   .1.   .0.
110   011   111   111   011   110   111   111   110   111   011   111
.1.   .1.   .01   10.   11.   .11   .1.   .1.   .1.   .0.   .10   01.
```

End point mask set preserving the following 4-connected situations in 2d.

```
110   001   000   000
010   011   010   110
000   000   011   100
```

Appendix C.

Test images as used in the simulation of the A* algorithm on the Path Finding Machine (PFM). From (Dekker 1987).

(Note that grey points are object points (constraints), black points are optimal path points, S = start point, G = goal point)

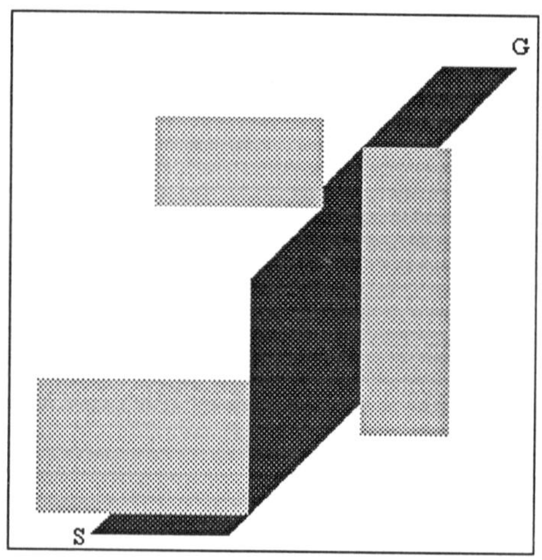

Image b: Simple constraint image (256 x 256); S = (46,10); G = (240,230)

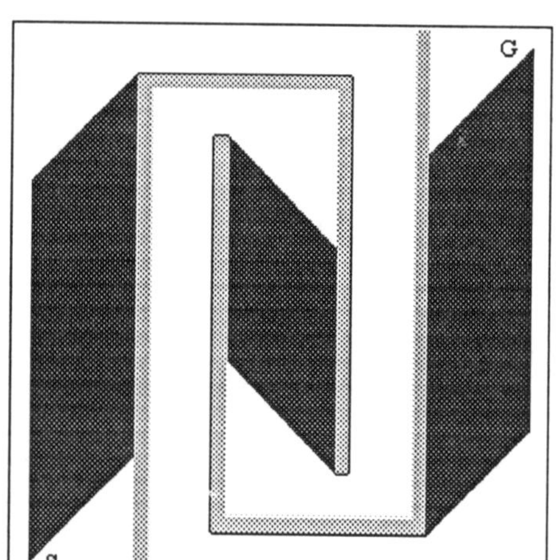

Image c: Simple 2D maze (256 x 256); S = (10,10); G = (245,245)

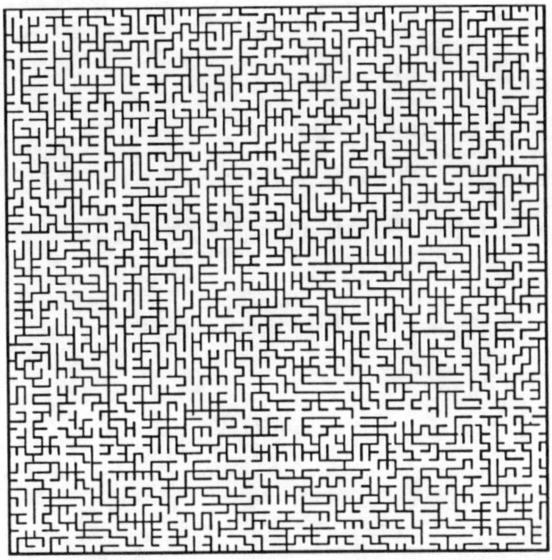

maze

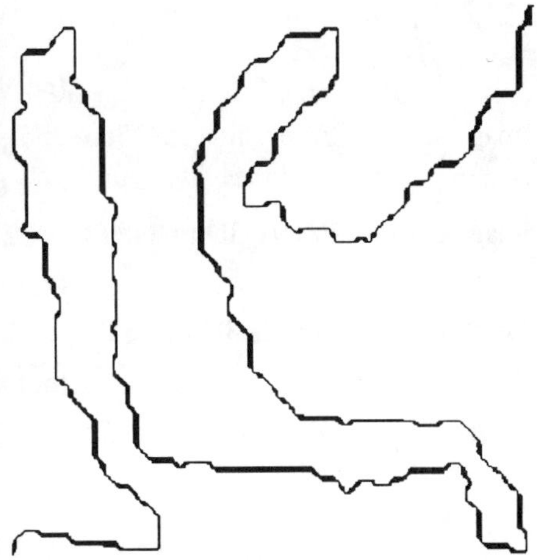

resulting path

Image d: Difficult 2D maze (256 x 256); S = (0,0); G = (254,254)

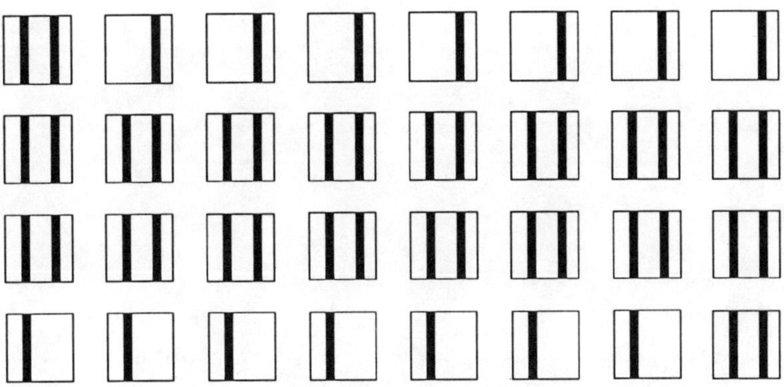

constraint image

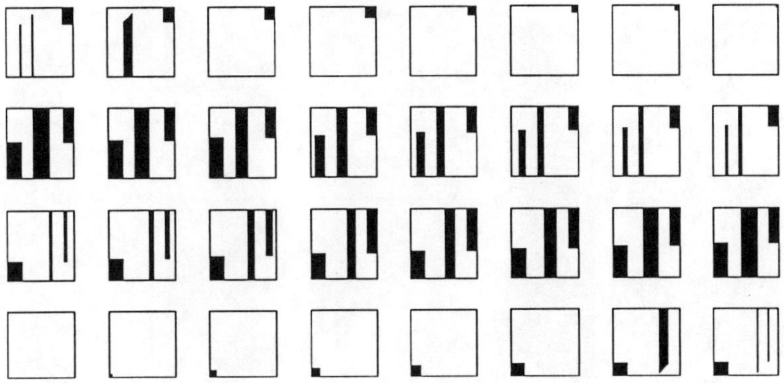

resulting path

Image f: Simple 3D constraint image (32x32x32); S = (0,0,0); G = (31,31,31)

(not shown)

Image a: Empty image (256 x 256); Start = (20,32); Goal = (246,224)

(not shown)

Image e: Empty 3D image (32 x 32 x 32); S = (2,0,0); G = (29,31,31)

(not shown)

Image g: Simple 3D maze (32 x 32 x 32); S = (3,0,1); G = (0,1,30)

(not shown)

Image h: Difficult maze (32 x 32 x 32); S = (0,0,0); G = (30,30,30)

Acknowledgements.

At this point I would like to express my gratitude to all those that were involved in the accomplishment of this thesis.

First I would like to express my deep gratitude to Bob Duin, who put me on the computer architecture track (again) and who has been and still is a very enthusiastic and stimulating companion, especially in digging deep to find the truth.

This research originally started as a comparative study between architectures, mainly focussed on the DIP-1 and the CLIP-4, a study that eventually was finished by Erwin Komen. At that time it was felt essential to develop a special chip with an architecture wholly different from that of the CLIP-4 but with the same capacity.
The desire to design and realize this processor chip (the CLPE) was enthusiastically welcomed and financially supported by Guus Schwippert, managing director of the Delft Center for Micro-Electronics. Although more skeptical but still confident, Jan van Zijverden, director of the Institute of Applied Physics-TNO, supported this. I appreciate their moral and financial support that was crucial at the start of the project.
The realization of the CLPE would not have been possible without the support of professor Patrick Dewilde and Rein Nouta of the Network Theory Group of the Faculty of Electrical Engineering. I appreciate the close and amicable cooperation with them. The background pushing by Rein Nouta to get the various chips realized (3!) has been especially valuable.
The final realization of the chip at Philips-Valvo in Hamburg would not have been possible without the wholehearted support of Ir. Engel Rosa at the Philips Physics Laboratory in Eindhoven (Natlab). I appreciate his open mind for university needs very much.
I am indebted to my first student (guitar) Martin Kraaijveld, who slaved at the computer for months to get the chip designed and simulated in Nmos, wondering if this all there was in life. No, but without that and him there would not have been a chip.
I am indebted also to his successors Jimmy Venema and, notably, Cees Schot, who realized the test chip and the final CLPE. I more and more realize that the character of a terrier is necessary to realize integrated hardware. I would like to thank Ronald de

Knijff, also blessed and cursed with this attitude, who performed the testing of the CLPE on the Genrad Test system of DIMES.

My warm appreciation goes to Sito Dekker, who has been the energetic motor and a stimulating discussion partner in the field of the dataflow processing and the path finding machine.
I thank Erwin Komen for his work on grey value processing on the CLPE and the way in which he finished the job I had left. As with Erwin's work, this work was partially supported by the Foundation for for Computer Science in the Netherlands (SION) with financial support from the Netherlands Organization for Scientific Research (NWO).

The work on thinning conditions would not have been as crystalized if Albert Vossepoel would not have been there to be my "censor" and discussion partner. I deeply appreciate his willingness to engage in this "battle" with me on a field that I do not consider my own. Again Bob Duin, and Piet Verbeek have my appreciation for the right remarks and questions on crucial moments.

I would like to thank the entire staff of the Pattern Recognition Group for their radiant "brotherhood of science and technique" climate. Especially Professor Ted Young, head of the group, who (although sometimes you would ...) has a remarkable charm and skill to keep malignant conflicts far away, thus creating an atmosphere in which research is a pleasant and exciting activity.

A special word for Willem van der Horst, for his cover design and hints on the lay-out and Caspar Williams "voluntary" courier to Deventer, hometown of Kluwer.

Without the numerous people that traversed my path and encouraged me on my road of science, I would not have been at this point.

Finally, my wife, Yolande, who explicitly did not want to be mentioned. Despite a continuous loyalty conflict between her love, her motherhood and her own career, she supported me to accomplish this work.

Curriculum Vitae.

Pieter Jonker was born in Amsterdam in 1951. After his HBS-b diploma at the St. Nicolaas Lyceum in Amsterdam, he studied Electrical Engineering at the Twente University of Technology. He received a Bachelors degree in 1977 and an Engineers degree in 1979 in the digital technique group of Prof. Dr. G. A. Blaauw.
From 1980 to 1982 he worked as a scientific researcher in the field of micro-electronics at the Central Laboratory of the TNO research organizations.
From 1982 to 1986 he worked on subjects in micro-electronics and Flexible Production Automation at the Institute of Applied Physics-TNO. He stated the architecture and supervised the realization of a Flexible Manufacturing System in co-operation with the Metal Research Institute-TNO.
Since 1985 he collaborated with Dr. Ir. R.P.W. Duin in research on the topic Pipelines versus Processor Arrays and in 1986 he was appointed assistant professor in the Pattern Recognition Group of Prof. Dr. I.T. Young at the Faculty of Applied Physics of the Delft University of Technology.
Since 1987 his research activities in Applied Physics range from computer architectures and VLSI circuit design for Image Processing to Robot vision and Flexible Assembly Systems. He supervised the diploma work of many students. Within the field of architectures, he pursued the development of a special integrated circuit for image processing (the CLPE) and researched and realized various small hardware architectures.
He initiated and supervises the research of a major multi-disciplinary project on Flexible Assembly, a cooperation between four faculties of the Delft University, which involves robot vision, robot path planning and collision avoidance, robot control, object manipulation, assembly process planning, production scheduling and error management.
He is member of the advisory board of the CIM Center Delft and participated in various small research projects for the Dutch industry. He is a member of board of the Dutch Institute for Silicon Based Signal Processing (COSIGN). He is lecturer of a PhD course on multi-dimensional computer architectures and he is on the teaching staff of the Special Course on Image Processing for Industrial Applications, organized by the (Dutch) Center for Image processing and Pattern recognition (CBP). He is currently chairman of Technical Committee 6, (Special Architectures) of the International Association for Pattern Recognition (IAPR).